RISKY REWARDS

Risky Rewards

How Company Bonuses Affect Safety

ANDREW HOPKINS AND SARAH MASLEN
The Australian National University, Australia

ASHGATE

Published by
Ashgate Publishing Limited
Wey Court East
Union Road
Farnham
Surrey, GU9 7PT
England

Ashgate Publishing Company
110 Cherry Street
Suite 3-1
Burlington, VT 05401-3818
USA

www.ashgate.com

British Library Cataloguing in Publication Data
A catalogue record for this book is available from the British Library

The Library of Congress has cataloged the printed edition as follows:
Hopkins, Andrew, 1945- author.
    Risky rewards : how company bonuses affect safety / by Andrew Hopkins and Sarah Maslen.
        pages cm
    Includes bibliographical references and index.
    ISBN 978-1-4724-4984-9 (hardback : alk. paper) -- ISBN 978-1-4724-4985-6 (ebook) -- ISBN 978-1-4724-4986-3 (epub)
    1. Industrial safety--Economic aspects. 2. Bonuses (Employee fringe benefits) 3. Incentives in industry. 4. Risk management. I. Maslen, Sarah, author. II. Title.

HD7262.H67 2015
658.4'08--dc23

2014029240

ISBN   9781472449849 (hbk)
ISBN   9781472449856 (ebk – PDF)
ISBN   9781472449863 (ebk – ePUB)

Printed in the United Kingdon by Henry Ling Limited, at the Dorset Press, Dorchester, DT1 1HD

Contents

List of Figures

List of Tables

Chapter 1
Introduction

Organisations often act in ways that seem irrational or contrary to their interests. In particular they sometimes fail to attend properly to hazards that lead predictably to disaster. A case in point is the petroleum company, BP, which has suffered catastrophic accidents twice in recent years: the Texas City refinery explosion in 2005, and the oil well blowout in the Gulf of Mexico in 2010. These events took a huge human toll: a total of 26 people died, hundreds were injured and countless friends and relatives were bereaved. But it is the economic cost which is relevant here. The Texas City explosion cost the company billions and significantly affected its share price, although it did not threaten the existence of the company. The oil well blowout cost BP even more dearly – in the order of $40 billion. Given that the market value of BP at the time was about $120 billion, this was a body blow that came close to destroying the company. Three years later the share price remained 25% below where it was at the time of the blowout. Given the scale of these economic consequences, BP's failure to devote more resources to controlling these hazards looks to be economically irrational, at least in hindsight.

Such irrationalities make a lot more sense, however, when we recognise that organisations themselves don't act – individuals within them do. Behaviour that seems irrational from an organisational point of view may be far more intelligible when seen from the point of view of individual actors. Their failure to spend money on the prevention of major accidents may indeed be quite rational for them. Major accidents are rare, and underinvestment can continue for years without giving rise to disaster. On the other hand, managers are judged on their annual performance, especially with respect to profit and loss. Consequently, spending money on the prevention of major accident events is not necessarily in their short-term interest. On the contrary, cutting expenditure on maintenance, supervision and training may enhance short-term profits, while inexorably increasing the risk of disaster in the longer term. Moreover, business unit leaders tend to think in the short term because they may only be in a particular management position for a couple of years before moving on. They may thus be long gone before the results of their cost-cutting decisions become apparent. At least one commentator, Bergin, has seen this as a root cause of the Texas City explosion: "Managers did not act to prevent Texas City (he says) because every incentive and potential penalty they faced told them not to."[1] The same could be said for the Gulf of Mexico blowout.[2]

In order to fully understand the way the incentive system worked to distract attention from major hazards at Texas City, and for that matter in the Gulf of Mexico, we need to develop the distinction between personal and process safety. This distinction was highlighted in all of the reports following the Texas City accident, but was given greatest prominence in the Baker Report, which expressed it as follows:

> *Personal* or *occupational* safety hazards give rise to incidents – such as slips, falls, and vehicle accidents – that primarily affect one individual worker for each occurrence. *Process* safety hazards give rise to major accidents involving the release of potentially dangerous materials, the release of energy (such as fires and explosions), or both. Process safety incidents can have catastrophic effects and can result in multiple injuries and fatalities, as well as substantial economic, property, and environmental damage. Process safety in a refinery involves the prevention of leaks, spills, equipment malfunctions, over-pressures, excessive temperatures, corrosion, metal fatigue, and other similar conditions.[3]

In some contexts, process safety hazards are referred to as major hazards or major accident hazards.[4] These latter terms are more general: they apply to rail and air transport and underground mining, all of which can experience catastrophic accidents, although they cannot be described as process industries.

Since personal safety and process safety are concerned with different kinds of hazards, it is logically possible to focus on one type of hazard and not the other. This was the essence of the problem at Texas City and in the Gulf of Mexico accident: the focus was on personal safety hazards, not process safety hazards. This was not a conscious choice. Many managers did not understand the distinction or, if they did, they assumed that attention to personal safety hazards would automatically ensure that attention was given to process safety.

Nor was this a personal failing on the part of individuals – it was the result of a deep-seated and widespread organisational failure. Safety is commonly measured using workforce injury statistics (e.g. lost-time injuries, recordable injuries, first aids). A low injury rate is arguably evidence that conventional occupational hazards are being well managed (to be discussed later), but such statistics imply nothing about how well *process* hazards are being managed. The problem is that catastrophic process incidents are by their nature rare, and even where process hazards are poorly managed, an installation may go for years without the sort of process safety incident that gives rise to multiple fatalities. So, if an organisation is seeking to drive down its injury rate, it will naturally focus on the hazards that are contributing to that rate on an annual basis. These may be vehicle hazards, working at heights, trip hazards and so on. The stronger this focus, the more likely it is that the organisation will become complacent with respect to major hazards, precisely because they do not contribute to the injury rate on an annual basis.

There is one major hazard industry that has not made this mistake – the airline industry. For the airline industry, major hazard risk refers to the risk of aircraft loss. No airline assumes that having good personal injury statistics implies anything about how well aircraft safety is being managed. The reason, no doubt, is that there is just too much at stake. When a passenger airliner crashes, hundreds of people are killed. The financial and reputational costs to the airline are enormous and there is the real risk that passenger boycotts might threaten the very existence of the business. Moreover, unlike those killed in industrial accidents, many of the victims of airliner crashes are likely to have been influential and/or to have influential relatives, which tends to magnify the costs and other consequences for the airline. For all these reasons, airlines have developed distinctive ways of managing aircraft safety and would never make the mistake of using workforce injury statistics as a measure of aircraft safety. It is just as senseless in process industries as it is in the airline industry to assume that injury statistics tell us anything about how well major hazards are being managed.

Consider now how BP's incentive system worked at Texas City, and in the Gulf of Mexico. Bonuses were paid to people at all levels of the company largely on the basis of productivity and cost minimisation. There was a safety component, but this was largely determined by injury statistics. There was nothing in the bonus structure to focus attention on how well major hazard risk was being managed. Texas City experienced numerous dangerous gas releases each year. This could easily have been treated as an indicator of how well process safety was being managed and included in bonus calculations. But it wasn't. It was this that led Bergin to conclude that "Managers did not act to prevent Texas City because every incentive and potential penalty they faced told them not to."

This problem was identified in various reports following the Texas City accident. The Baker panel report made the following recommendations:[5]

> A significant proportion of total compensation of refining line managers and supervisors [should be] contingent on satisfactorily meeting process safety performance indicators and goals ...

> A significant proportion of the variable pay plan for non-managerial workers ... [should be] contingent on satisfactorily meeting process safety objectives.

These seem like eminently sensible suggestions. And they have been very influential. In recent years many companies in process industries have recognised that the criticisms levelled at BP applied to them as well, and various companies, including BP, now include some indicators of process safety in their bonus structures. The basis on which BP paid its annual bonus to its most senior executives in 2012 is presented in Figure 1.1. The value placed on rebuilding trust is evidently a response to the Gulf of Mexico disaster. Other than that, the main point to note at this stage is the safety component which, at 30%, is the highest we have seen. Total recordable injury frequency is a measure of personal safety, while loss of containment events – leaks and spills – is a measure of process safety. This is a bonus scheme that, if anything, gives greater weight to process safety than to personal safety. In principle, this will focus company attention on major hazard and environmental risks far more effectively than previously.

Safety and risk management (30%).

  – Recordable injury frequency.

  – Loss of primary containment.

  – Process safety-related major incident announcements and high-potential incidents.

Rebuilding trust (20%).

  – External reputation.

  – Internal morale and alignment.

Value creation (50%).

  – Operating cash flow.

  – Underlying replacement cost profit.

  – Total cash costs.

  – Gearing.

  – Divestments.

  – Upstream production efficiency.

  – Upstream major project delivery.

  – Refining and Marketing net income per barrel.

Figure 1.1   Bonus structure for top BP executives, 2012[6]

A more cautious approach

The preceding account ends on an optimistic note. But before drawing any firm conclusions about these new developments, there are a number of questions we need to ask about whether incentive systems really function as intended. First and perhaps most fundamentally, *are* we primarily driven by financial incentives, or are we motivated by other kinds of rewards, such as job satisfaction and positive feedback from supervisors, that make financial rewards largely irrelevant? Second, assuming that there are circumstances in which we respond to financial rewards, do these rewards motivate the intended behaviour or do they simply encourage people to manipulate the measures ("game the system" in contemporary jargon) in such a way as to gain the desired reward without achieving the intended ends? In short, do they have unintended and perverse consequences that undermine them? Third, are the indicators that are being used to measure catastrophic risk the right ones? It is incumbent on those who advocate incentive systems to consider these questions carefully and to examine critically the new incentive systems that companies are developing to identify their strengths and weaknesses. This book starts, therefore, with general considerations about human motivation and incentive systems and moves ultimately to an empirical study of particular incentive arrangements in a number of companies operating in industries where there is the potential for disaster. These arrangements turn out to be surprisingly complex. Accordingly, a significant part of the book is devoted to examining the logic of these schemes and the impact they might be expected to have.

Our empirical study is based on 11 companies operating in hazardous industries. We examined documents provided by these companies and carried out multiple interviews with the senior people in each company most able to describe how their systems worked. The companies were drawn from the oil and gas, petrochemical and mining industries. Clearly there are other industries with the potential for catastrophe, for example, aviation, rail, nuclear power and even the finance industry, as the global financial crisis demonstrated. Our study is therefore limited to a subset of industries in which catastrophe can occur. The companies were mostly large multinational concerns, with headquarters in several different countries. A couple of the companies selected were smaller, operating in a single country. Most were publicly listed companies, although a few were wholly owned subsidiaries of larger, multinational operations. Another limitation of the study is that we have no data on national oil companies (NOCs), which are influential players on the world stage. The incentive systems that operate in NOCs may be systematically different from those identified here and a follow-up study on this subset of major hazard firms would be of considerable interest. The companies selected are not a statistically random sample; rather, the sample is purposive: designed to cover the range of possibilities in the industries concerned. Studying these firms therefore gives us a picture of the problems and possible solutions. Finally, we interviewed a number of senior managers in a subset of these companies about just how they responded to these incentive schemes.[7]

Outline of book

The book falls naturally into two parts. The first part, consisting of Chapters 2 and 3, sets out the issues in more detail, while the second part, consisting of Chapters 4, 5, 6 and 7 sets out our empirical findings in relation to some of these issues.

In Chapter 2 we discuss some research findings on the impact of incentives on human behaviour and their relevance in the present context. The chapter highlights the work of Daniel Pink who argues that in this matter there is "a mismatch between what science knows and what business does".[8] In particular, he says, science knows that people are motivated more by intrinsic rewards (e.g. job satisfaction) than by extrinsic rewards (e.g. money), yet business continues to pay financial bonuses as if economic motivations were all that mattered. Pink has been very influential in his advocacy, to such an extent that at least one company we have studied has decided to abandon aspects of its bonus system and rely on 'conversations' initiated by the CEO and propagated down through the organisation. The expectation is that these conversations will engage people's intrinsic motivations and that this will ensure alignment with company objectives more effectively than paying bonuses. Pink's argument is a serious challenge to the line of argument we developed earlier in relation to BP and in particular calls into question the Baker panel recommendation that companies should aim to include process safety in their bonus arrangements.

However, as we shall see, the findings on which Pink relies do not apply in any straightforward way to the issue of company bonuses. We can begin to get a glimpse of the difficulty by remembering that bonuses are often paid in the context of performance evaluation. This means that the size of the bonus reflects the opinion of the evaluator on how well the person has performed. In this context a large bonus amounts to a congratulatory handshake while a small bonus is inevitably read as a criticism. In this way the bonus becomes a psychological reward as well as an economic one. Much of the research on the impact of monetary rewards pays no attention to the psychological component of bonuses as actually paid in the corporate world, making the research of dubious relevance, if not totally irrelevant, in the present context.

An important aspect of this question of the impact of incentives is the issue of unintended or perverse consequences. Chapter 2 deals with this as well. It begins by noting that this problem besets policy makers in many fields, such as health care and education. When it comes to safety, one of the unintended consequences of paying bonuses for low injury rates has been the underreporting of incidents. Underreporting may stem directly from pressure not to report. It may also result from sophisticated classification strategies that ensure that injuries do not count as injuries for the purpose of bonus calculations. The problem is so great that various authorities have argued that bonuses should never be paid for low injury rates. The same problem applies to process safety indicators which are likely to be underreported or manipulated in the same way as injury data, if it is at all possible to do so.

Chapter 3 deals with the problem of paying bonuses to prevent rare but catastrophic events. The problem here is that where bonuses are paid on the basis of annual performance there may be nothing in the incentive arrangements to ensure proper attention is being paid to the prevention of rare events. The chapter identifies four ways to deal with this.

The first is to delay for up to ten years the payment of a bonus earned in any one year, and to make the final payment dependent on performance during the ten-year deferment period. Long deferrals like this were recommended for the finance industry in the wake of the global financial crisis. At least in theory it gives top executives a vital interest in longer-term catastrophic risk management. Whether it does in practice depends on the detail of such schemes. The second way to encourage a focus on catastrophic risk is to identify indicators of how well major hazard risk is being managed in the here and now, and to pay bonuses in relation to these indicators rather than on the basis of whether or not a major accident actually occurs. The third way is to identify actions that the most senior managers can take to improve safety and include them in the performance agreements of these managers. The fourth is to acknowledge and reward people who take initiatives in relation to major hazard risk, in particular by making incident reports that help the organisation avoid disaster.

Chapter 4 is the first of our "findings" chapters. It takes up the issue of deferred bonuses paid to the most senior people in large corporations. When we examine the detail of these schemes we find that they encourage top managers to focus on maximising financial return over the period of deferment, *even if this means running a greater risk of experiencing a catastrophic event*! Long-term bonuses are intended to align the interests of top managers with the interests of shareholders. Many shareholders and certainly many share beneficiaries, such as pension fund members, seek security rather than profit maximisation. We conclude that current long-term bonus schemes are not in their interest.

Chapter 5 examines annual bonuses – their design and potential effect. Two factors which go into the determination of these bonuses are the performance of the group, in most cases the global corporation, and the performance of the individual. The people at the very top of the corporation have the capacity to influence the performance of the corporation in a perceptible way. Rewarding them for group performance therefore has some potential to motivate their behaviour. But for the rest, it makes little sense to reward them on the basis of group performance, since there is nothing they can do to affect this performance in a perceptible way. On the other hand, the individual component of the bonus is based on a supervisor's evaluation of how well the individual met the objectives specified in his or her performance agreement. This has obvious motivational potential. However this potential is undermined by the requirement that most people be evaluated as average (or words to that effect), which means that most people feel damned with faint praise. One company in our sample has explicitly recognised this problem and modified its bonus scheme accordingly.

Chapter 5 also looks at the extent to which safety, both personal and process safety, are included in these bonus calculations. We find that personal safety has a relatively high profile in the determination of group performance, but that process safety with few notable exceptions is almost invisible. When it comes to individual performance, safety, both personal and process, is relatively inconspicuous, although again there are some very important exceptions.

Chapter 6 reports on what our interviewees say about the motivational effects of their bonuses, in particular their annual bonuses. Many regarded the performance agreement as little more than a restatement of the job requirements for the year. That being so, what really shapes their priorities is not the performance agreement itself, but the performance review with their supervisor. This interview gives them the clearest indication of what is really important to their supervisor and how best to please him or her. Moreover, according to respondents, what they seek most is positive feedback – an indication from their supervisor that their contribution is valued. In many cases the monetary reward is seen as symbolising this evaluation rather than being an end in itself. For managers at the top, the situation is a little different. Their performance agreements contain numerical targets and these provide a clear indication of priorities. Even so, it remains a moot point as to whether it is the financial reward or the positive feedback in the evaluation process that is of most significance.

In Chapter 7 we turn our attention to indicators of major hazard risk. Many companies are developing such indicators, although they do not yet feature prominently in bonus arrangements. This chapter critiques the logic of the indicators in use and suggests ways in which they can be improved.

Finally, Chapter 8 is not merely a summary of what we have found. We make some quite harsh judgments about the systems we studied and we propose a number of ways in which bonus arrangements can be better focused on major hazard risk.

We are immensely grateful to the companies that allowed us to examine them under the microscope, and we hope that, in so far as they are able to recognise themselves, they will take our criticisms as constructive.

Notes

1 Bergin, 2011, p. 85.
2 Hopkins, 2012.
3 *The Report of the BP US Refineries Independent Safety Review Panel* (Baker Report), Washington, US; Chemical Safety and Hazard Investigation Board, *Investigation Report: Refinery Explosion and Fire*, January 2007, p. x.
4 For example, in jurisdictions that require safety cases.
5 Baker Report, op. cit., p. 251.
6 http://www.bp.com/assets/bp_internet/globalbp/globalbp_uk_english/set_branch/STAGING/common_assets/downloads/pdf/IC_DRR11_directors_remuneration_report_2011.pdf.
7 Our study was funded in part by the Australian Energy Pipelines Cooperative Research Centre (EPCRC). This is a joint industry/university/government organisation that conducts research relevant to the oil and gas pipeline industry. It is part of the Australian Government's Cooperative Research Centres Program. The cash and in-kind support from the Australian Pipeline Industry Association Research and Standards Committee is gratefully acknowledged. The present study is part of a broader research program, funded by the EPCRC, on the sociological aspects of major accidents and their prevention. For more details of the research methodology, see Appendix 1 of this book.
8 Pink, 2009. Quotation taken from the cover.

Chapter 2
Questioning the Value of Financial Incentives

The whole strategy of using financial incentives to achieve business objectives has been called into question by certain contemporary writers, most prominently, Daniel Pink.[1] In his book, *Drive*, Pink assembles a mass of data that purportedly demonstrate the failure of the financial incentive strategy. He concludes that:

> The problem is that most businesses haven't caught up to this new understanding of what motivates us. Too many organisations – not just companies, but governments and nonprofits as well – still operate from assumptions about human potential and individual performance that are outdated, unexamined, and rooted more in folklore than in science. They continue to pursue practices such as short-term incentive plans and pay-for-performance schemes in the face of mounting evidence that such measures usually don't work and often do harm.[2]

Pink has acquired international guru status. His books are best sellers and companies are acting on his advice. It therefore behoves us to examine his arguments carefully. But rather than making Pink and his conclusions the central focus of this chapter, it will be convenient to identify three questions that Pink implicitly raises in *Drive* and to treat these as the framework for the discussion. The questions are:

- Are we primarily motivated by financial incentives, or are other, non-financial motivations more important?

Assuming financial incentives do have some effect,

- Do they destroy other motivations?
- Do they generate perverse consequences that render them counterproductive?

A focus on these questions enables us to go beyond the mere critique of Pink and to set the scene more effectively for the rest of this book.

Are we motivated by financial incentives?

The first of the above questions is way more complex than might at first appear. To answer it we look first at the issue of human motivation, and second, at how this applies to bonuses as actually paid in the corporate sector.

Human motivation

There is a spectrum of views about human motivation. At one end, neo-classical economic theory and more recently public choice theory[3] start from an assumption that people are rational, self-interested calculators and therefore respond in predictable ways to financial incentives. At the other end, sociology, psychology and more recently behavioural economics argue that real human beings are driven by a variety of motives, in addition to self-interest. Sociology, in particular, holds that these alternative motives are much more significant than economic self-interest. In principle, therefore, these latter disciplines call into question the efficacy of financial incentives. The neo-classical paradigm is so well known that it needs no further description here. In the following paragraphs we deal first with sociology, touch briefly on psychology and move finally to behavioural economics.

The discipline of sociology began as a reaction to the idea that the rational self-interested individual is the indivisible building block of which society is constructed. The sociological view is that humans are constructed by the societies or smaller-scale communities in which they are embedded. Their beliefs, values and motives vary, depending on this social context, and are irreducibly social. So, for example, if modern western individuals act to maximise financial advantage, it is because they are embedded in a capitalist society that encourages them to behave in this way. On the other hand, in feudal society, glory and honour were often more important than wealth, and people in certain social classes were willing to die for these ends, as did so many European knights who went off on the crusades. The modern day equivalent of this motivational structure can be seen in the jihadists and suicide bombers who emerge from or are inspired by certain traditional Islamic societies, which in some respects are the feudal societies of today. These holy warriors, both mediaeval and modern day, behave in ways that are incomprehensible to classical economics but are entirely explicable in terms of the norms and values of the societies from which they come.

It is in capitalist society, that economic self-interest achieves its greatest prominence as a motive, but it is worth observing that capitalism itself will not flourish unless located in a certain kind of society, where people fundamentally trust each other, or more generally, where there is some minimal level of social capital. It has been demonstrated, for example, that northern Italy flourished economically in the second half of the twentieth century because of relatively high levels of social capital, while in the south, the dominance of the extended family made it difficult to trust people outside the family circle and hence to do business with them.[4]

Leaving aside the social origins of human motivation, sociological research has found that humans are driven by a variety of motives such as, reputation, fairness, duty, curiosity, creativity and altruism, and that these are in fact more important in understanding human behaviour than purely financial or self-interested motives.[5] Even more strikingly, whatever the goal, economic or otherwise, people do not typically select the most rational means to achieve that goal; their choice of means is influenced by values and emotions.[6]

These ideas are fundamental in psychology. To give just one example, consider Maslow's theory of the hierarchy of needs. Maslow at first identified five needs, starting with the most basic, *biological and physiological needs*, number 1 in Figure 2.1, and culminating in self-actualisation. In 1970 he added three more levels, resulting in the eight levels depicted in Figure 2.1.

Maslow's theory challenges the view that humans are rat-like pleasure seekers. Even if not correct in all its details, it calls into question any corporate strategy that relies purely on financial incentives to motivate people.

---

8. *Transcendence needs* – helping others to achieve self actualization.

7. *Self-Actualization needs* – realizing personal potential, self-fulfillment, seeking personal growth and peak experiences.

6. *Aesthetic needs* – appreciation and search for beauty, balance, form, etc.

5. *Cognitive needs* – knowledge, meaning, etc.

4. **Esteem needs** – self-esteem, achievement, mastery, independence, status, dominance, prestige, managerial responsibility, etc.

3. *Social Needs* – Belongingness and Love, – work group, family, affection, relationships, etc.

2. *Safety needs* – protection from elements, security, order, law, limits, stability, etc.

1. *Biological and Physiological needs* – air, food, drink, shelter, warmth, sex, sleep, etc.

Figure 2.1   Maslow's hierarchy of needs

In recent years the sociological critique of neoclassical economics has been popularised by behavioural economics.[7] One of the most famous experiments in behavioural economics is the so-called "ultimatum game". Two players are given a sum of money, say ten dollars, which is to be divided between the two, provided they can agree on the division. Player 1 proposes a division. Player 2 then decides whether to accept the offer. If the offer is declined, neither player gets anything. Suppose player 1 proposes to keep nine dollars and give one dollar to player 2. Will player 2 accept? The choice for player 2 is either one dollar or nothing. If player 2 acts in a rational, self-interested way, s/he will accept. Experiments show however that player 2 typically rejects the offer in these circumstances, because the division is perceived to be unfair. This is just one of numerous pieces of data assembled by behavioural economists that demonstrate that people will act contrary to their economic interest where other motivations and values are at stake.

What can we conclude from this discussion? First, the theory of financial incentives rests on an impoverished conception of human motivation. Secondly, for this reason, it cannot be assumed that financial incentives will determine behaviour. Third, as a corollary, the question of whether financial incentives motivate us in particular circumstances is an empirical one and is likely to depend critically on the circumstances.

Pink is in agreement with all of these conclusions. In particular, he accepts the fact that in some circumstances financial incentives do indeed motivate behaviour. Most notoriously, the Global Financial Crisis (GFC) of 2008 was precipitated by people whose behaviour was a direct consequence of the regime of financial incentives in which they operated. The subprime mortgage brokers were paid on the basis of the volume of loans they could generate, and many brokers grew rich quickly by luring people into borrowing money they could not afford. When the bubble broke, there was financial pain all around, except for the mortgage brokers who retained their original commissions. This was a remuneration system that encouraged brokers to focus on the very short term and gave them no incentive to take account of the almost inevitable longer-term consequences.[8] In other words, the GFC provides a stark example of people responding to financial incentives and hence, incidentally, a stark example of the need to ensure that these incentives do not have unintended consequences, a matter we return to later.

Pink makes a similar point in relation to company executives. For many companies,

> quarterly earnings are an obsession. Executives devote substantial resources to making sure the earnings come out just right. And they spend considerable time and brainpower offering guidance to stock analysts so that the market knows what to expect and therefore responds favourably. This laser focus on a narrow, near-term slice of corporate performance is understandable. It's a rational response to stock markets that reward or punish tiny blips in those numbers, which, in turn, affect executives' compensation.[9]

Complementing this picture, at the bottom of the corporate hierarchy, there is evidence that for routine tasks that involve only mechanical skill, the larger the incentive the greater the response.[10] Piece rate systems of pay are based on this very premise.

This is not evidence that financial incentives *invariably* serve to motivate people. It is evidence simply that there are *some* circumstances in which they do. However it does call into question some of Pink's more sweeping conclusions, such as his claim that most businesses haven't caught up with the new understanding of what motivates us.

Bonuses for corporate employees

If the motivational effects of financial incentives are dependent on the circumstances, we need to examine those circumstances more closely. The time has come, therefore, to introduce some of the empirical findings of our own work. In the large companies we have studied, bonus payments are made to people at many levels in the hierarchy. The schemes are complex and depend on a number of factors, which can be divided into two categories.

The first category consists of measures of group performance, the group being the company, or perhaps a business unit within the company, or even some particular operational group with the business unit, such as a drilling crew. The measures might relate to such things as profit, production and injury rate, all in relation to the particular group.

The second set of factors relates to the performance of individuals. In many companies, people at professional and managerial levels have written performance agreements with their supervisors and they are assessed individually as to how well they have achieved the goals set out in the performance agreement. This individual assessment then contributes to their bonus. Waged workers typically do not have such individual agreements, and their bonus payments, if they receive them, are determined entirely by the performance of some relevant group, such as the business unit.

Consider now the type of motivation that is accessed by these bonus arrangements. First, group-level incentives. Most people would perceive that their behaviour can have no discernible effect on most group-level measures, especially if the group is large, such as the company itself or one of its business units. Furthermore, if people cannot influence the measures, then bonuses cannot function as financial incentives, since what any one individual does makes no discernible difference to the bonus. However, where the group is small enough that individuals feel they can make a difference, as in the case of a drilling rig, the bonus to the group does become a direct financial incentive to the individual.

This is not to say that group-level bonuses are generally ineffective. While most members of the relevant group cannot influence group-level measures, group leaders can. That after all is what they are employed to do. Group-level bonuses therefore provide group leaders with a personal, financial interest in improving the production figures, as well as improving performance in relation to other metrics included in bonus calculations, such as injury rates. Furthermore, leaders create culture by what they attend to and what they prioritise[11] and such cultures percolate down through the organisation ("what interests my boss fascinates me"). Where the lead from the top is sufficiently insistent, it will end up affecting the culture of the work group, that is, "the way we (the workers) do things around here". This is obvious enough in the case of production pressures, but there is plenty of evidence that where top leaders use a range of carrot-and-stick options to inculcate certain safety-relevant behaviours (e.g. fastening seat belts, wearing hard hats), these behaviours also become part of the culture of the work group.

Now here's the interesting point. Culture is not just about the way things are in fact done around here. There is also a normative component. Subtly, the way we do things is transformed into a belief about the way we *ought* to do things. And this normative component of the culture in turn influences the behaviour of individuals, regardless of whether or not there is any discernible financial advantage. People end up taking pride in achieving the goals of the business group, precisely because those goals have become the goals of the work group. These mechanisms are clearest in relation to production goals – work groups often take on the production goals of the organisation in this way – but they are equally applicable to other goals such as injury rate reduction. The net effect is that leaders, who may themselves be financially motivated, end up creating cultures that exert influence on lower-level managers and workers, independently of any financial incentives that may be paid.

Consider now the second factor that affects the bonus payments of people in the corporate world – personal evaluation by their supervisors. These evaluations are based on the supervisor's judgement about how well the individual has performed in relation to the various tasks specified in the performance agreement. Typically this judgement is qualitative, rather than quantitative. For example, if one of the specified tasks is to develop a procedure for something, the important question is not, has the procedure been written, but how good is the procedure. This is inherently a qualitative and subjective judgement.

The overall evaluation of an individual is often made on a four- or five-point scale such as:

outstanding;
exceeds expectations;
meets expectations;
below expectations.

Furthermore, in an attempt to reduce the level of subjectivity, supervisors are expected to provide a certain proportion of assessments in each category. A typical requirement would be that at least 70% of evaluations must lie in a central category ("average", or "meets expectations" in the scale above), with perhaps only 5% rated as outstanding or exceptional. There is, finally, some pre-determined formula by which these rankings are converted into bonus payments.

This system of performance evaluation can have powerful effects on people's behaviour. One of the inquiries into the blowout in the Gulf of Mexico found that individual performance agreements often specified that employees should contribute to the cost reduction goals. The inquiry found further that of 13 employees whose evaluations it examined, 12 had documented ways in which they had saved the company large sums of money. One had put together a spread sheet showing how he had saved the company $490,000![12] Clearly the incentive arrangements were having the intended effect.

But what motivations are these incentive schemes tapping? Presumably, in part, financial benefit is a direct motivator. But there is much more to it. The rating on the four- or five-point scale is a judgment by the supervisor of how well the individual is performing. The rating involves praise, or faint praise, or criticism by the supervisor. Powerful psychological incentives are therefore at work. People will seek to please their boss in order to achieve these psychological rewards, independently of any material rewards involved.

Let us now relate this discussion of bonus arrangements for corporate employees to the theoretical considerations outlined earlier. That discussion was about the extent to which we are motivated by non-material considerations such as values and the opinions of others, as opposed to economic self-interest. It concluded that we are motivated by many things and that there are circumstances in which non-financial goals are the most important. However, this does not automatically undermine the strategy of paying financial incentives to corporate employees, because, curiously perhaps, these financial incentives do not rely for their effect on economic self-interest alone. Instead they tap a number of human motives, among them the need for approval, the need to belong and the need to be recognised as making a valuable contribution, all higher level motives in Maslow's hierarchy of needs that transcend purely economic considerations. In short, the fact that human motivation is complex, and in particular more complex than neo-classical economic theory assumes, does not in itself undermine the potential of financial bonuses to influence behaviour in corporate settings.

Do bonuses destroy other motivations?

The second question that Pink raises is whether incentive payments destroy other motivations we may have, such as creativity and curiosity. As we shall see, this is a significant issue for educators seeking to motivate student learning, but its relevance in the employment context is less clear. After all, an employer might well take the view that, provided a financial bonus is having the intended effect, its impact for other types of motivation is unimportant. Let us leave this reservation aside for the moment and see where Pink's question leads.

The starting point is a series of experiments dating back to 1969 that show that extrinsic rewards (money, recognition) can dampen intrinsic motivation (e.g. curiosity, creativity). One such experiment can be fairly easily described. Researchers identified a group of pre-school children who tended to spend their free time drawing. They then divided these children into two groups.[13]

Each was given paper and felt-tipped pens and invited to spend time drawing. One group was told that if they drew something they would receive a "good player" certificate with their name inscribed and a ribbon attached. The other group was offered no such inducement. Two weeks later, the teachers provided paper and pens for use during a free play period. The children who had previously been in the no-reward group drew with just as much relish as they had done originally. The children who had previously been rewarded, but were not now being offered any reward, showed much less interest and spent much less time drawing.[14]

There have been numerous such experiments leading to the overall conclusion that "tangible rewards tend to have a negative effect on intrinsic motivation".[15] Pink puts it strikingly: "Try to encourage a kid to learn math by paying her for each workbook page she completes – and she'll almost certainly become more diligent in the short term and lose interest in math in the long term."[16] However the research demonstrates that this effect is *far less* for college students than for school-age children.[17] We return to this in a moment.

It is interesting that in the research described above the reward offered was not material; rather it was symbolic – the symbolic recognition that the child was a "good player". In this research tradition, material and symbolic rewards are grouped together under the single heading – tangible.

As distinct from tangible rewards, *verbal* rewards tend to *enhance* intrinsic motivation. According to the research, the effect is variable, but where the verbal reward amounts to positive feedback that enhances the individual's perceived competence, this enhances intrinsic motivation.[18]

The implication would seem to be that businesses should rely on verbal rewards and do away with all systems that provide either material or symbolic rewards for good performance, on the grounds that they dampen intrinsic motivation. However it will be recalled that the negative effect of tangible rewards on intrinsic motivation were far less for college students than for children. This suggests that for adults in the world of work the effect may be even less. As noted earlier, this research has greatest relevance in the field of education, but its implications in the world of work are less clear.

Furthermore, recall that the effect on intrinsic motivation in the experiment was noted *when the external reward was no longer on offer*. In the world of work, external rewards remain on offer and will continue to engage extrinsic motivations, regardless of what the effect on intrinsic motivation may be.

There is one aspect of the research findings that suggest that the payment of bonuses on the basis of performance evaluations will not necessarily undermine intrinsic motivation. Recall that performance evaluations are about how well people have done their agreed job. A good evaluation amounts to positive feedback and the educational research suggests that positive feedback will enhance intrinsic motivation. (The problem noted before is that because of the requirement to grade on a curve, relatively few people are given this positive feedback. The system in other words does not take full advantage of its potential to motivate people.)

Pink's own conclusions from this work on extrinsic and intrinsic motivation is more far-reaching. He argues that businesses need to redesign themselves so as to take advantage of intrinsic motivation. More concretely this means they must design work so as to maximise job satisfaction. The job characteristics that make for greatest job satisfaction, he says, are autonomy (freedom to work on one's own terms and in one's own time), mastery (the ability to develop one's skills and abilities and to apply them to new challenges) and purpose (the possibility of making a contribution to the world). If companies design work so as to allow these motivations to come into play, there will be no need for additional extrinsic motivators, apart of course from some reasonable level of fixed pay. *Drive* is largely about how organisations can make this move to more satisfying work.

Pink stresses that this is not a utopian vision. He identifies various companies that have redesigned work in this way. However it has to be said that these companies are largely in the business of knowledge creation, the bulk of the staff being software developers, designers and others doing high-level creative work.[19] It is harder, but not impossible[20] to implement Pink's recommendations in industries that produce manufactured goods or energy. For such industries, his vision might best be seen as a long-term or aspirational goal.

If we focus again on the fact that in the corporate world bonuses are paid on the basis of performance evaluations, there is another reason to doubt the relevance of the pre-school drawing experiment and others like it. Performance evaluations are based on how well the employee has performed in relation to what are often quite nebulous goals, such as contributing to cost savings. Here are the goals specified in a performance agreement we have seen for a drilling engineer.

- deliver wells without exceeding budget;
- work with rig to develop cost or efficiency savings ideas;
- work with vendors on developing lower cost alternatives;
- deliver continuous improvement in performance in drilling speed.

Such goals do not deprive employees of autonomy; the second and third, in particular, challenge them to think imaginatively and to use their skills to a particular corporate end. Arguably, bonuses paid in these circumstances, whether tangible or verbal, encourage rather than crush curiosity and creativity.[21]

This last conclusion is consistent with other research about the impact of extrinsic incentives on creativity. The research compared the level of creativity that artists displayed in works they were paid to produce with the level of creativity they displayed in their non-commissioned work. Creativity was judged by a panel of experts who were blind as to the purpose of the research. The findings were that when the commission specified the task tightly, the level of creativity dropped, but when the commission gave the artists a relatively free rein, the level of creativity was not negatively affected. The researchers therefore distinguished between two types of extrinsic rewards: controlling and enabling. Where rewards are controlling, creativity is diminished; where rewards are enabling, creativity is unaffected.[22]

The implication of this research for performance bonuses is clear. Where the performance goal leaves plenty of room for initiative in deciding just how to achieve that goal there is no reason to think that financial rewards will dampen intrinsic motivations such as curiosity and creativity. Interestingly, Pink agrees: "The science shows that it is possible – though tricky –

to incorporate rewards into non-routine, more creative settings without causing a cascade of damage."[23]

Do bonuses generate perverse effects that render them counterproductive?

This third question is the most serious challenge of all to the use of financial incentives, especially, as we shall see, for safety. The problem is that wherever a goal is expressed in terms of some numerical measure, this can encourage attempts to manage the measure, independently of the phenomenon or activity being measured. This is the problem of perverse or unintended effects. We begin this discussion by demonstrating just how widespread this problem is.

Experience elsewhere

Consider health care. Governments in some countries have attempted to improve health services by introducing a variety of performance measures and making funding contingent on good performance against these measures. This is a situation in which perverse consequences can be confidently predicted. Research on the English National Health Service (NHS) identified 20 different types of unintended consequences that have occurred. We shall not summarise them here but simply give an example. In an effort to reduce emergency waiting times in hospitals the NHS introduced a five-minute emergency waiting target. Hospitals had to report their performance against this target. This "led to the employment in some hospitals of 'hello nurses'. They merely made contact with the patient in the first five minutes in order to 'tick the box' … This was costly in resources and resulted in little benefit for patients."[24]

A second cautionary tale comes from UK attempts to assess university research output:

> The first Research Assessment Exercise used as its main indicator the number of refereed research publications. The result was a proliferation of new journals, and the growth of undesirable practices, such as the publication of essentially the same work in different guises in different journals and the splitting up of research papers into several smaller ones. Also the pressure to get published disadvantaged long-term research.[25]

A further consequence was that academics shied away from publishing in professional and popular journals, because these were not refereed. The result was a reduction in the dissemination of knowledge to the wider community.

Finally, university teaching suffered. Researchers put less effort into teaching, and less time, preferring to employ teaching assistants to carry out their teaching functions. The use of indicators to provide incentives for one kind of university activity (research) had the effect of reducing the quality of another university activity (teaching) for which there were no corresponding incentives. We put it this way to facilitate a comparison with company bonuses. The message from Texas City with which we started this book was that the incentives to maximise production and also to maximise personal safety enhanced performance in these areas, while the absence of any incentives for process safety meant that performance in this area went steadily downhill.

A third example of the problem concerns the payment of bonuses to senior public servants. A former top Australian public servant, Allan Hawke, has written persuasively about the negative consequences of bonuses in his experience.

> Where such schemes are in operation, most individuals or groups will work towards optimising their performance, regardless of its effect on the system. Creative accounting, goal displacement, withholding information, reduced quality at the expense of more output, individual visibility which discourages cooperation and other gaming strategies are the perverse results of such a perverse system.[26]

Hawke argued that one of the most destructive features of the system he experienced was that it required people's rankings to be distributed on a bell-shaped curve. Under these circumstances,

> half of your people will learn that they are below average – a statistical inevitability. Some may accept their fate; others will view this as proof that their manager is incompetent. Some will redouble their efforts to prove the judgment and system wrong – that may be noticed, and they may be lucky enough to be ranked above average next time – if so, someone above average last time will fill their below average slot this time.[27]

And sarcastically he writes,

All of this must do wonders for morale and superior workplace performance.[28]

The problem of forcing a bell-shaped distribution was alluded to earlier. Companies that use a four- or five-point evaluation scale are not condemning approximately half their staff to a below average evaluation. But they are evaluating a substantial majority as being merely average performers, which is almost as disheartening. Hawke has a solution to this. He advocates that organisations pay a salary increment each year, in the normal course of events, and the performance evaluation process be limited to deciding whether this normal increment is justified.[29] For the great majority of people it will be, and the performance evaluation will amount to a form of positive feedback, not a deflating judgement that one's performance is merely average.

Experience in the business sector

Consider now the perverse or unintended consequences of bonus payments in the business sector, starting with bonuses that aim to enhance economic performance. First, they can encourage a focus on short-term economic gain at the expense of longer-term economic consequences, as happened in the GFC. Second, they can encourage totally unethical behaviour. Pink provides an example. A company that set targets for "its auto repair staff (found that) workers responded by overcharging customers and completing unnecessary repairs".[30]

Third, and particularly relevant in the present context, bonuses paid to managers for cost cutting can result in cuts to maintenance, training and supervisory staff, all factors that have been implicated in major accidents.[31] This is not an unethical response, since those making these cuts do not necessarily appreciate the possible consequences. The increased risk of major accident is simply an unintended consequence of a cost-focused bonus system.

Consider now the use of bonuses as a means of reducing injury rates. The perverse consequences of such bonuses are widely acknowledged. One of the most common measures of injury is the lost-time injury rate. A lost-time injury is an injury that results in time away from work other than the day on which the injury occurred. If people are brought back to work the day after an accident and placed on alternative duties, hey presto, a potential lost-time injury is no longer a lost-time injury. While return to work can often be justified from an injury management point of view, there is plenty of anecdotal evidence of people being brought back to work purely as a means of managing the measure.[32] Many industries have sought to overcome this particular problem by focusing on broader injury categories, such as all injuries requiring any medical treatment or even first aid. But these, too, are measures that can be manipulated. A bandage or plaster is first aid only if it is administered by a nurse; if it is self-applied it does not count. There is evidence that companies encourage self-application for this reason.[33]

The US Occupational Safety and Health Administration (OSHA) is very critical of the payment of bonuses based on injury rates because of the potential of such schemes to suppress reporting.[34] Employees know that if they report an injury they may be affecting not only their own bonus, but that of their workmates. This puts enormous pressure on people not to report.[35]

OSHA is particularly critical of incentives that are paid for achieving zero injuries, for example, where "a team of employees is awarded a bonus if no one from the team is injured over some time period". The pressure to not report in this situation is overwhelming.

OSHA goes on to list other ways in which bonuses could be used "to encourage safe work practice, in particular, incentives that promote worker-participation in safety-related activities, such as identifying hazards or participating in investigations of injuries, incidents or 'near misses'".

There are, however, some important refinements that need to be made to OSHA's argument. Two different situations can be distinguished. The first is where the group for which the accident rate is calculated is small enough that a single incident will make a difference. It is precisely in these situations that the incentive for ordinary workers not to report their injuries is greatest, and OSHA's argument applies with greatest force. A detailed example of this is provided in the addendum to this chapter. It is very unwise to provide ordinary employees with injury rate bonuses in these circumstances. The second situation is where the group for which the rate is calculated is so large that a single injury will make no discernible difference, for example, if the rate in question is for the company as a whole. In this case, bonuses paid to ordinary employees provide no incentive to suppress reporting, but neither do they provide any incentive to be more careful. To include injury rates in bonuses in such circumstances is undesirable: at best it is superfluous and at worst may encourage cynicism.

These conclusions do not apply to the managers who are responsible for the relevant groups. For them, the bonus functions not as an incentive to avoid injuring themselves, but as an incentive to reduce the number of injuries experienced by others. This will probably require changes to job design, equipment and procedures, even behaviour modification programs. All this will require additional resources, and the bonus provides higher-level managers with an incentive to find these resources.

Of course, such a bonus also provides an incentive for group managers to suppress reporting by others, if they can. But the more remote these managers are from potential reporters – both physically and organisationally – the more difficult it is for them to suppress reporting, even if they were inclined to do so.

This suggests a way in which the perverse effects of injury rate bonuses might be avoided. Suppose the CEO and top managers are rewarded for reducing injury rates, but no one further down the hierarchy is so rewarded. This means that regardless of the group size for which injury rates are calculated, there will be no incentive for ordinary workers to suppress reporting. The system thus has a better chance of operating as intended. Admittedly, the mere fact that the CEO and top executives have a financial interest in reducing the number of reported injuries may create a culture that discourages reporting. To counteract any such tendency it should be possible to incentivise ordinary employees to report accidents when they happen, for example, by making small bonus payments for filing reports correctly. The strategy in others words would be to provide top managers with an incentive to drive accident rates down while at the same time providing ordinary employees with an incentive to ensure that the reported rate reflected the true rate as accurately as possible. We shall not develop this further here, although it is obvious that such a scheme would need careful thought, and perhaps experimentation, to minimise the potential for further unintended outcomes.

These cautionary comments about the use of injury rate data are equally applicable to process safety indicators such as loss of containment events. The US standard on process safety indicators (API 754)[36] defines a Tier 1 event as one that involves a gas release of 500 kg or more per hour, while a Tier 2 event is a gas release of between 50 and 500 kg per hour (there are corresponding figures for petrol and oil releases). However it is not easy to determine the quantity of gas released. Weight must be calculated from estimates of pressure, size of hole and so on. The fact that the figures are only estimates leaves plenty of room for data manipulation in order to ensure that the figure does not cross some threshold. If the estimate comes in at 495 kg then it is only a tier 2 event, not a tier 1, and such a number has almost certainly been adjusted to achieve this outcome.

Such perverse outcomes are ubiquitous and bedevil most systems of bonus payments that operate in the private sector. Pink is right to draw attention to them. Bonus arrangements need to be carefully designed to guard against these perverse consequences. If companies are not willing and able to give this the attention it deserves, the evidence is that financial bonuses should be abandoned.

Conclusion

Writers such as Pink have challenged the payment of bonuses in the corporate world on at least three grounds: (i) financial bonuses ignore other human motivations that may be more powerful than economic self-interest; (ii) financial incentives destroy other motivations, and (iii) financial incentives create unintended and perverse outcomes. However much of the evidence for these claims is only tangentially relevant to the financial bonuses actually paid to employees in the corporate world. In particular, the fact that bonuses are not just financial rewards but also a form of recognition means that much of the research on the incentive effects of bonus payments is irrelevant. The most serious issue that Pink raises is the third, that of unintended consequences. These are rife, and companies need to be continuously vigilant in this regard, since even the best-designed systems are likely to have such consequences.

Addendum to Chapter 2

This addendum provides a case study that demonstrates the potential for unintended consequences to creep into even carefully designed incentive schemes. The scheme was developed for drilling operations on land. It offered a performance bonus of $200, for every person in the drilling crew, for every day the team was able to shave off the estimated number of days required to drill a particular well. The estimate for any given well was based on time taken to drill comparable wells. So for, example, if the estimated time was 25 days but drillers were about to cut five days off this time, the performance bonus paid for this well was $1000 per person. This is a significant production incentive. Clearly it needs to be matched by equally significant safety incentives, so that drillers do not seek to maximise their bonuses at the expense of safety. Accordingly, there was a safety and environment bonus of $50 per person per day of drilling, provided the well is completed without significant incident. So, in the above example where the well took 20 days to drill, each person stood to receive a safety and environmental bonus of $1000, by chance in this case, the same as the performance bonus. But beyond this, the system was designed to make safety the overriding consideration, because in the event of a recordable injury, *neither* the environmental and safety bonus, *nor* the production bonus would be paid for that well.

This last proviso puts enormous pressure on injured workers to not report the injury. Not only is their own bonus at stake but so too is the bonus of all team members. One can expect that creative ways will be found to avoid reporting in this situation. Interestingly, the company has considered this possibility and added another proviso in the policy – that failure to report a safety or environmental incident may lead to the termination of the safety and environmental bonus program (but not the termination of the production bonus).

This policy represents a sustained attempt to deal with the perverse consequences of many bonus arrangements. The bonus is designed to ensure that production does not take precedence over safety. And it is designed to deal with the incentive not to report. But there are still foreseeable problems. The incentive not to report injuries is so great that workers are likely to find ways to avoid reporting, despite the risk this poses to the bonus system itself. Moreover when someone is injured, it is an accidental event for those involved, accidental in the sense of unintended, unforeseen and regretted. Yet all concerned are penalised severely and the message conveyed is that all the other efforts that may have been made by the team to drill the well quickly and safely count for nothing in these circumstances. This is likely to generate a sense of injustice with undesirable consequences for group morale.

One other thing to note about this system is that it seems bound to create disillusionment in the longer term. If teams consistently perform better than average, the average itself will come down, so it will be harder and harder to achieve production bonuses. In the end people may end up feeling punished for their good performance. All in all, a sad tale of how unintended consequences can creep into even quite well thought out schemes.

Another company we spoke to explained how it had introduced a bonus arrangement much like the one described above, but had abandoned it, because of all these unintended consequences.

Notes

1 See also Flemming, 2011.
2 Pink, 2009, p. 9.
3 See Etzioni, 1988, pp. 56ff.
4 Putnam, 1993.
5 Nor can these non-economic values be unified with economic considerations in a single all-embracing concept of utility. To do so renders the concept of utility tautologous. If it is impossible to conceive of non-utility maximising behaviour, the theory that behaviour is utility-maximising loses all predictive power. See Etzioni, 1988, pp. 25–31.
6 Evidence to support these general claims will be found in Etzioni, 1988.
7 Ariely, 2009.
8 Yeh, 2010, p. 112.
9 Pink, 2009, p. 55.
10 See references in Pink, 2009, p. 60.

11 Schein, 1997, p. 231.

12 DWI [Deepwater Investigation], transcript from the joint BOEMRE/Coast-Guard inquiry, originally available at www.deepwaterinvestigation.com, 7 October 2010, p. 148.

13 Actually three. The description here is a simplified version of the experiment.

14 Pink, 2009, p. 36.

15 Ibid., p. 37.

16 Ibid.

17 Deci et al., 2001, p. 14.

18 Ibid., p. 3.

19 Pink, 2009, p. 85.

20 There is a massive literature on job enrichment.

21 These four goals show a heavy emphasis on cost and can be expected to undermine safety, unless balanced by a similar set of safety-related goals. This is a crucial point but it is not relevant to the argument here which is simply that providing financial incentives to achieve these goals does not necessarily undermine creativity.

22 Amabile, 1996, pp. 119 and 175.

23 Pink, 2009, p. 63.

24 Mannion and Braithwaite, 2012. See also, Goddard et al., 2002; Werner and Asch, 2005.

25 Elton, 2000, p. 276.

26 Hawke, A., Performance management and the performance pay paradox, 2011, unpublished paper, p. 19.

27 Ibid.

28 Op. cit., p. 20.

29 Hawke, op. cit., p. 18.

30 Pink, 2009, p. 49.

31 Hopkins, 2008.

32 For example, here is what one worker wrote in a culture survey at the Texas City Refinery in 2004: "Auto accident en route to sister site – unavoidable – but we were treated as if we caused the accident even though the other driver was cited. Forced to come to work via taxi when unable to drive and under pain medication which causes drowsiness – bad headache – unable to perform job; management was trying to avoid a lost-time from work. Personal concern was felt, but more concern was given to avoid lost time from work!" Hopkins, 2008, p. 86.

33 Here is an account provided by a worker in the culture survey mentioned in the previous note: "minor steam burn resulting in first aid visit; management encouraged self-treatment to avoid OSHA recordable injury!" Ibid.

34 https://www.osha.gov/as/opa/whistleblowermemo.html.

35 This bonus-related effect is sometimes compounded, OSHA notes, by employer policies of taking disciplinary action against employees who have accidents.

36 "Process Safety Performance Indicators for the Refining and Petrochemical Industries", STANDARD 754 published by American Petroleum Institute.

Chapter 3
The Problem of Rare but Catastrophic Events

The main focus of this book is the use of bonuses to promote the proper management of catastrophic risks. As we made clear in the introduction, the problem is that where bonuses are paid on the basis of annual performance there may be nothing in the incentive arrangements to ensure proper attention is being paid to the prevention of rare but catastrophic events. This chapter discusses various ways to deal with this.

The first strategy is to delay the payment of a bonus for several years and to make the final payment dependent on company fortunes during the deferment period. This gives people a vested interest in longer-term performance, and in particular on preventing major accidents that might impact on company fortunes. A second strategy is to identify indicators of how well major hazard risk is *currently* being managed and pay bonuses in relation to these indicators, rather than on the basis of whether or not a major accident actually occurs. A third approach is to identify things that senior managers can do to reduce catastrophic risk and include these in their performance agreements. Fourth, companies can reward outstanding examples of initiatives taken by individuals to reduce the risk of catastrophic events. The most important initiative in this respect is the reporting of bad news.

Deferred bonuses

Bonuses paid for financial performance generally outweigh those paid for safety performance. For the top leaders this imbalance is often overwhelming, with the result that their attention is focused far more on financial than safety performance.

However a major accident can have a major impact on company finances, as BP's recent experience demonstrates. It is worth documenting just how great this impact was before we go on to examine the implications. In 2005/6 BP suffered three safety and environmental failures that cost it dearly. First, the Texas City Refinery explosion in March 2005 cost BP some $2 billion in compensation to victims and another $1 billion in overhaul of the site.[1] Second, at the end of 2005 BP's revolutionary and only partially completed deepwater production platform in the Gulf of Mexico, Thunder Horse, suffered a structural collapse and tipped sideways. Fortunately no one was killed, but production was delayed by years and repairs cost $100 million. Finally, a few months later in March 2006, oil was discovered leaking from a BP pipeline in Alaska. The leak was caused by pipeline corrosion that had not been repaired or even identified. Evidence of further severe corrosion was then uncovered, leading to a shutdown of Prudhoe Bay, the largest oil field in the US.

Collectively, these events had a substantial effect on BP share price. From the start of 2005 to mid-2007, BP shares underperformed the European oil and gas sector by 16%. More broadly, while Amex oil index rose 72% from March 2005 to December 2007, BP shares rose only 15%.[2] In short BP paid a significant financial price for its failure to manage effectively its safety and environmental risks.

Even more dramatic was the effect of the Gulf of Mexico blowout. It has been estimated that the cost of this disaster to BP was of the order of $40 billion, about a third of the company's market value at that time. Shares lost half their value and three years later were still 25% below their pre-blowout level[3] (see Figure 3.1).

BP's experience demonstrates that even where bonus structures are overwhelmingly focused on company financial performance, all those who receive such bonuses have a potential interest in preventing catastrophic events. The issue to be addressed here is how that potential interest can be converted into a real and motivating interest. A useful starting point for this discussion is the experience of the finance industry.

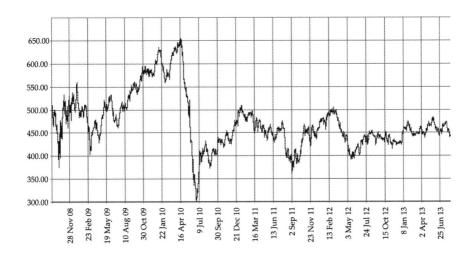

Figure 3.1   BP share price before and after Gulf of Mexico blowout, 20 April 2010[4]

The finance industry

The Global Financial Crisis of late 2008 was generated by years of risky behaviour in the finance industry. These risks had paid off for myriads of decision makers in the short term, but the long-term consequence was financial catastrophe. Since 2008, a number of ideas have emerged about how to alter the incentive arrangements for finance industry decision makers in such a way as to make them more concerned about the longer-term consequences of their behaviour. Most relevant here are the suggestions to do with shares and share options. Strange as it may seem, companies can hold shares in themselves (a truly Alice in Wonderland idea if one thinks about it for a moment). The result is that company-owned shares can be used to remunerate employees. This is now a standard form of remuneration for senior executives. Such systems are generally designed to give senior executives an interest in the financial performance over the longer term, not just the year in which the remuneration is earned. This is achieved by withholding the benefit earned in any one year, until a later year, and making the value of the benefit dependent on the share price in that later year. There are at least two ways this can be done – first, the share option. Employees who are to be remunerated at time 1 can be given the option to buy company-owned shares at time 2, at the price that prevailed at time 1. If the market price goes up between times 1 and 2, the employee can buy the company-owned shares at the lower price and immediately sell them on the open market for the higher price. If the market price has gone down in the interim, the share options are worthless. A second way is to award a certain number of company-owned shares at time 1, but to withhold them until time 2. If the share price has gone up in the interim, the value of the awarded shares is correspondingly greater. If the share price has decreased, the value of the award is correspondingly reduced. For ease of exposition in what follows we talk about the first of these methods – the share option – but the points are equally applicable to deferred share awards and to other similar schemes.[5]

The share option strategy aims to ensure that the most senior executives of the company take decisions that maximise shareholder value. However there are various problems with the strategy. In particular, senior executives may manipulate the share price in such a way as to maximise the value of their shares at the time they are seeking to receive their entitlements. In so doing they may be making decisions that are not consistent with the longer-term interests of the company and its shareholders. As other commentators have put it, "executives may enter into transactions that improve the current bottom line but create large latent risks that could cripple the firm in the future".[6] Yeh provides the example of Lehman Brothers to demonstrate this kind of behaviour. Lehman Brothers, which collapsed in late 2008, had transferred its risky investments to another company controlled by Lehman. Lehman was still exposed but this was not obvious to shareholders. The company looked financially healthier than it was and its share price remained higher than it would otherwise have been.[7] Whilst there is no evidence in the Lehman case that the behaviour was motivated by a desire to maximise the value of share options, it certainly had that effect. Moreover there is good evidence that, in general, senior officers actively manipulate share prices in order to maximise the value of their share options.[8]

The Director of the Program on Corporate Governance at Harvard Law School, Lucian Bebchuk, together with various colleagues, has proposed a system that would curtail the ability of senior executives to choose the time at which they exercise options and sell the shares so acquired. They propose, first, that executives should not be free to exercise their options in the year they are earned but should be required to wait for a period of years, perhaps seven years. Even if they leave the firm they would still need to wait the seven-year period. This would give them an interest in the long-term financial well-being of the firm. A number of firms have already adopted this model, some going so far as to prevent executives from receiving their entitlements until they retire.[9] But even this is inadequate, Bebchuk argues. Long-serving executives, perhaps at the point of retirement, would still have an incentive to manipulate the market price by means of various short-term strategies so as to maximise the value of their shares at the time of receipt. To remove this incentive, Bebchuk's second recommendation is that schemes be designed so that executives can only cash out their options gradually, over a period of years.

A far more radical idea has been floated by Stuart Yeh. He suggests that all remuneration above $100,000 and 20% of remuneration below $100,000 should be held back, by law, and converted to share options. The employee would be entitled to cash out 20% of the options in each successive year after they were earned, so that the full value of options would not be realised for five years. Remuneration earned each year would be deferred in this way.

For firms that are not traded on the stock exchange, Yeh's suggestion is that they be required by law to hold the delayed remuneration in a capital fund that would be drawn on to pay out breach of duty claims against the firm. Such a mechanism would be analogous to the capital accounts that law firm partners maintain in order to satisfy claims against their partnerships. It would give all employees an incentive to act prudently and also to ensure that fellow employees are acting prudently. As Yeh says, "delayed compensation would create powerful incentives for each professional to limit risky behaviour, and to report the risky behaviour of other individuals, so that the consequences of that behaviour do not imperil the payout of each professional's compensation".[10]

Hazardous industries

These ideas have direct relevance in the present context. Many companies in hazardous industries offer two types of financial incentive: short term and long term. Short-term incentives are tied to annual group performance and to individual performance against goals specified in performance agreements, as discussed in Chapter 2. The long-term incentives are determined in a two-stage process. Each year eligible people receive a certain number of shares or share options. The exact number will be based on a variety of factors. Thereafter the value of the shares/ share options is determined exclusively by the behaviour of the share price during the years between when the bonus is earned and when it is actually provided to the recipient. The lessons from the finance industry are directly applicable to these longer-term bonuses and, indeed, many have already been implemented. One that has not is the proposal to defer the right to exercise share options for seven years. This could be adopted with minimal modification. It is a device that, on the face of it, encourages the most senior executives to act prudently with respect to catastrophic risk, whether that risk be financial, safety or environmental. Assuming for the moment that this proposal were to be adopted, one modification we would suggest is that the figure of seven years be replaced by ten. Our reasoning is as follows. BP had been engaged in savage cost cutting at Texas City from the moment it acquired the site in 1999. From this point of view the explosion of 2005 was six years in the making. However, a 2002 review noted that the site had serious "integrity and reliability issues" that were "clearly linked to the reduction in maintenance spend of the last decade".[11] From this point of view the latency period was substantially longer. We conclude that it may take longer for short-sighted cost cutting to generate catastrophic safety outcomes than it does for short-sighted lending strategies to unravel catastrophically. Hence if anything the deferral period should be even longer than has been proposed for the finance industry. In keeping with this, one of our interviewees suggested that the appropriate deferral period might be up to ten years.

However these are provisional ideas only. The actual incentive effects of deferred bonuses will depend on the precise design of long-term bonus schemes. Our findings in Chapter 4 will require us to modify these provisional suggestions.

Bonuses for the management of catastrophic risk

The deferred bonus strategy implicitly ties bonuses to the *outcomes* of the catastrophic risk management process, that is, to whether or not it is successful in preventing disaster. It is also possible to make bonuses dependent on how well catastrophic risk is being managed *in the normal course of events*, regardless of whether a catastrophic event actually occurs. Our focus in this section therefore reverts to *short*-term incentives which aim to reward people for their on-going activity.

The strategy here is to identify indicators of how well major hazard risk is being managed. Once these are identified they can be included in bonus arrangements in such a way as to motivate people to attend to major hazard risks. There is, however, a fundamental problem, namely, that major accidents are few and far between, and it is difficult to demonstrate empirically that whatever indicators we choose really do measure variations in major hazard risk. The situation is not hopeless, though, because the basic and widely accepted model of major accident causation, the so-called Swiss cheese model, provides a logical way to relate indicators to risk. We elaborate this in what follows.

The prevention of major accidents depends on defence-in-depth, that is, a series of barriers or defences to keep hazards under control. It is important to understand that the concept of barrier is not restricted to physical barrier, but includes non-physical barriers such as training, procedures, testing and engineering controls. Accidents occur when all these barriers fail simultaneously. The Swiss cheese model developed by Professor Jim Reason conveys this idea (see Figure 3.2[12]). Each slice of cheese represents a fallible barrier and accidents only occur when all the holes line up.

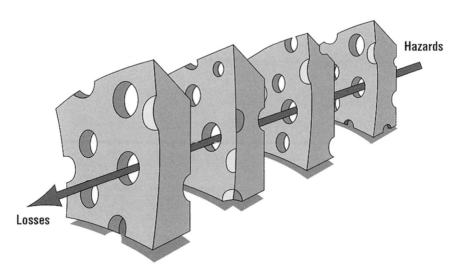

Figure 3.2   The Swiss cheese model

Precursor events

This model leads naturally to the idea of a precursor event, by which we mean an undesired event in which most barriers have failed and a major accident was prevented by the small number of remaining barriers, perhaps even a single remaining barrier. Take the situation that exists in many petrochemical facilities. The facility may be processing a highly flammable substance under high pressure. There will be a whole series of defences designed to keep the substance properly contained in pipes and vessels. Those defences include various aspects of the way the facility is designed, operated and maintained. But occasionally, if all these defences fail, the facility may experience a loss of containment. If it is a large loss, of either a flammable gas or a volatile and flammable liquid, the only remaining barrier to a fire or explosion may be ignition control policy which aims to minimise the potential ignition sources in the environment (for example minimising the number of vehicles on site). Clearly, the greater the number of loss of containments, the greater the likelihood of a major accident event. Conversely, the risk of such a major accident can be reduced by reducing the number of loss of containment events. This may be difficult to demonstrate statistically, but the logic itself is persuasive. A loss of containment is a *precursor* to an explosion at a petrochemical site in the sense that many major accident events at such sites begin with a loss of containment, although not all loss of containment events progress to become major accidents.

Texas City refinery was experiencing hundreds of loss of containment events annually prior to the 2005 explosion. Indeed the number had increased by 50% in the two preceding years, a clear indicator of the decreasing effectiveness of the process safety management system. Many petrochemical facilities have relatively few losses of containment, but as we aggregate data for facilities and business units, at some point the number of loss of containment events becomes large enough to be able to talk meaningfully of a loss of containment *rate*. This rate can then be taken as an indicator of major hazard risk, to be driven downwards.

The American Petroleum Institute has embraced this idea and promulgated a recommended practice, API 754, that treats loss of containments as a process safety indicator for the refining and petrochemical industries (see Figure 3.3). Confining attention to flammable gas releases, the standard defines a release of 500 kg

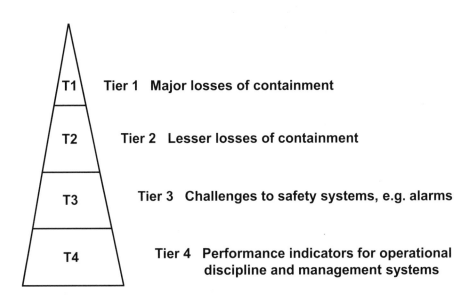

Figure 3.3   Schematic version of API process safety triangle

or more in any one hour as a Tier 1 process safety event, while a release of between 50 and 500 kg is a Tier 2 event.[13] Many companies round the world have now adopted API 754 and are using the numbers of Tier 1 and Tier 2 loss of containment events as indicators of how well their process safety management systems are functioning.

This important initiative is a direct response to the BP Texas City Refinery disaster and the inquiry into that disaster by the US Chemical Safety Board. The Board recommended to API that it develop a standard for "performance indicators for process safety in the refinery and petrochemical industries". The standard is thus limited in its scope; it does not apply to all major hazard operations. It specifically excludes the following:

- truck or rail operations, except on-site truck loading or discharge operations;
- releases from pipeline transfer operations occurring outside the process or storage facility fence line;
- marine transport operations, except where vessel is connected ... (for) product transfer;
- retail service stations;
- Petroleum industry drilling operations (part of *upstream* operations) are also beyond the scope of the standard, which is explicitly for the *downstream* segment of the industry (refining).

There is an important rationale for these exclusions. In the case of refineries and petrochemical facilities, loss of containment events are precursors to major accidents, in the sense discussed earlier. *In other contexts, loss of containment may NOT be a precursor to a major accident and conversely, the major accident events of greatest concern may have different precursors.* This is a crucial point which we develop in what follows.

Consider the road transport of flammable petroleum products, the first of the exclusions listed above. The major accident event of greatest concern in this context may be a road tanker accident that results in the rupture of the tank and the spillage and ignition of a large quantity of flammable material. It is hard to imagine loss of containment events that can sensibly be regarded as precursors to such an accident. However a road tanker accident can be thought of as a loss of *control* event. Other situations in which the driver lost control or came uncomfortably close to losing control of the vehicle are therefore the relevant precursors. One indicator that drivers are not in control of the situation to the desired extent is the fact they need to brake hard. Thus, hard braking can be seen as a precursor to a road tanker accident. In-vehicle-monitoring-systems can detect instances of hard braking, so it is feasible to treat the rate of hard braking as an indicator of risk in this context. Companies can then set about training their drivers to drive more conservatively, which should reduce the need for hard braking and hence the risk of road tanker accident. At least one company we know of has been persuaded by this logic and implemented such a scheme.

A second exclusion in the above list is marine transport operations. The big risks here are groundings, collisions between tankers, and possibly worst of all, large vessels colliding with production platforms, all of which can result in catastrophic fires or spills. It makes little sense to treat other loss of containment events as precursors to such accidents. More relevant are the occasions on which vessels deviate from authorised routes or find themselves on a collision course. Such precursor events can be readily identified with modern GPS systems and reducing the rate of such events would clearly reduce the risk of major marine accidents.[14]

There is an important precedent for this in the airline industry. A mid-air collision is one of the worst imaginable aviation accidents, and air traffic control is largely about preventing such accidents. A relevant precursor event is the failure of two aircraft to maintain an appropriate distance from each other. This is called a breakdown of separation. It does not necessarily mean that the two aircraft are dangerously close; merely that they are closer than they should be. That is the undesired precursor event. Accordingly, the frequency of breakdown of separation is one of the most important risk indicators used by air traffic control organisations around the world.[15]

A third exclusion from API 754 is the drilling industry. The most feared major accident event in this industry is the blowout. However loss of containment, as normally understood, is not a precursor to a blowout. When the idea of loss of containment is raised with drilling rig personnel, what first comes to mind is a leak from a hydraulic hose, which is certainly not a precursor to a major accident. A leak is of course a potentially reportable *environmental* event, if it escapes into the sea, and for this reason, rigs are concerned about what are, from the point of view of major accident prevention, quite trivial losses of containment.[16]

The most relevant precursor to the blowout in the drilling industry is the kick. Consider the following passage from an introductory text:

> Before a well blows out, it kicks. A kick is the entry of enough formation fluids [oil and gas] into the well bore so… [as to create an upward] pressure in the well. If crew members fail to recognize that the well has kicked and do not take proper steps to control it [by sealing the well], it can blow out … The key to preventing blowouts is recognizing and controlling kicks before they become blowouts.[17]

A kick, then, is a precursor to a blowout. Accordingly, the fewer the number of kicks, the less the risk of blowout. In the drilling industry the strategy must be to count the number of kicks and to drive this number downwards.[18]

A final exclusion from API 754 is the pipeline industry. There should be no assumption, therefore, that loss of containment is a relevant indicator of major hazard risk for pipelines. We will comment on this in more detail in Chapter 7.

It is important to highlight these exclusions from API 754. While for many major hazard industries, loss of containment will indeed be an important indicator of major hazard risk, for many others, and indeed for segments of the petroleum industry, it will not be the most relevant or useful indicator. Each industry and industry segment must therefore work out for itself what are its most feared major accident events and identify precursor events that can be counted and converted into an indicator of risk to be driven downwards.

Many petrochemical companies are now making bonus payments dependent on a reduction in the numbers of loss of containment events, as defined in API 754. But even where loss of containment is indeed a precursor to a major accident event, there is another question that needs to be considered before too much emphasis is placed on this indicator. This is the question of the manipulability. As discussed in Chapter 2, as soon as an indicator becomes important, we can expect attempts to manage the indicator rather than to manage the risk itself. API 754 categorises losses of containment in terms of the weight[19] of the substance that has escaped. These weights cannot be known exactly and must be estimated. Anecdotal information is that the engineers who make these estimates frequently adjust the assumptions on which the estimates are based in order to minimise the significance of releases. It seems likely that estimated release weights will be underestimates, wherever this is a matter that impacts on bonuses. Furthermore, some releases will only come to light if they are reported. Including such releases in bonus arrangements therefore provides a potential incentive not to report. Companies that include loss of containment numbers in bonus systems need to consider carefully these perverse consequences and put in place strategies to counteract them. This matter will be taken up later in the chapter.

Indicators of barrier health

The Swiss cheese model provides us with a second quite distinct way to develop indicators of how well major hazard risk is being managed. Rather than focus on precursor events in which several defences may have failed, the strategy is to measure the effectiveness or health of each barrier, separately, and to use this to build up a picture of the health of the system of barriers as a whole. This approach has been well expounded in a document published by the UK Health and Safety executive (HSE), entitled, "Developing Process Safety Indicators". The essence of the approach is the barrier performance indicator model (see Figure 3.4).

"RCS" in the diagram stands for Risk Control System, meaning barrier or defence, in the language of the Swiss cheese model. Notice that the controls are not simply physical defences; in Figure 3.4 they are all organisational: plant change (i.e. management of change); inspection and maintenance; compliance; permit to work.

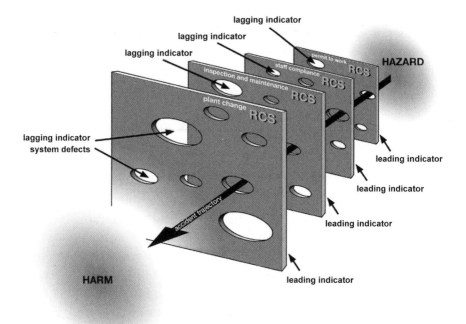

Figure 3.4   The UK HSE barrier performance indicator model

The HSE document identifies two types of indicator in relation to each of these defences:

A. Measures of routine, safety-related activity, e.g. proportion of safety critical instrument/alarm tests done on schedule;
B. Measures of failures discovered during routine safety activity, e.g. proportion of safety critical instruments/alarms that fail during testing.

The document calls these leading and lagging indicators respectively, but we shall not adopt this terminology here because it is ultimately confusing.[20] The HSE document suggests that a type A and type B indicator be chosen for each critical risk system. If monitored regularly, these will provide a good indication of the health of the system.

The document also identifies a third type of indicator, in addition to the A-B couplet:

C. Measures of control failure revealed by an unexpected incident, e.g. numbers of plant trips or critical alarms. These are an indication that prior process control has failed.

This scheme of A, B and C type indicators can be mapped back onto the lower half of the API 754 process safety triangle, although the mapping is not exact.

Type C measures correspond to Tier 3 which is defined as an event that "represents a challenge to the barrier system that progressed along the path to harm but is stopped short of a Tier 1 or Tier 2 loss of containment".[21]

Type A measures correspond to Tier 4 – measures of process safety management system activities.

Type B – failures revealed in the routine testing – do not fit neatly into either tier 3 or 4, as defined in the standard. However the standard treats them in practice as falling into tier 3.

This discussion of Types A, B and C indicators in the HSE document and Tier 3 and 4 indicators in the API standard is helpful in understanding the range of possible indicators and how they relate to barrier failure. However in practice most companies that have tried to include indicators of barrier effectiveness have not bothered with these niceties but have simply made use of whatever data is most readily available. This means that the barrier health monitoring strategy has remained only partially implemented to date.

This idea of using indicators of barrier health to drive improvement in major hazard risk management is evidently a good one – it provides a level of detail which enables real improvements to be made. However there are several difficulties that must be noted.

The first is that this approach involves a proliferation of indicators and hence a dilution of focus on any one. Senior managers understandably respond by asking for a summary version. One strategy that is often used is the traffic light scorecard. Indicators are arrayed on a spread sheet and categorised as red, amber or green, depending on their values. The number of reds, ambers and greens is then a rough-and-ready summary indicator of the health of the system. The danger is that this system becomes an end in itself and that senior managers will want to see as much green as possible on the spread sheet without asking how realistic this is. One leading petroleum company has explicitly recognised this problem and adopted the slogan – "challenge the green and embrace the red", meaning, be sceptical of the green, and welcome the red because it helps the company to identify where it needs to concentrate its efforts. On this view, too little red is actually a bad thing. We discuss this further in Chapter 7.

A second issue is that Type A (Tier 4) indicators – indicators of routine safety system activity – are particularly likely to generate unintended consequences as soon as they become indicators that matter. Take an indicator such as number of audit recommendations "closed out" (completed) on time. The purpose of such an indicator is clear enough. But if close-out rates are made to affect bonuses, managers may increase the *quantity* of close-outs by sacrificing their *quality*. As a result they may be able to meet whatever close-out targets are set, without necessarily improving safety. This is not dishonesty; simply pragmatism. It is a reasonable hypothesis that the more causally remote the indicator is from the outcome of concern, the greater the opportunities for perverse outcomes of this nature.

Here is the experience of one company in this regard, expressed in the language of leading and lagging indicators.

> In previous years a number of leading indicators have been introduced to the scorecard in an effort to drive improved performance. Typically these have been scored as green at the end of the year, whilst the lagging indicators (primarily injury rate) have often remained red. There is an underlying concern that a focus on such leading indicators has sometimes had the unintended consequence of driving short-term KPI delivery at the expense of systematic and sustainable implementation of robust HSE management systems and performance.

In short, for this company the indicators of safety management system activity bore little relationship to the real outcomes of the safety management system. Such are the pitfalls of indicators of safety system activity when they become indicators that affect remuneration.

Among the many potential indicators of barrier health some stand out as self-evidently related to major hazard risk. The following three examples are worth mentioning.

1. Number of approved deviations from engineering technical practices (the fewer the better).

There are two reasons why engineering authorities might approve deviations. The first is the standards are not in fact appropriate in the circumstances, perhaps because of new technology. If deviations are being approved for this reason then the use of the proposed indicator will drive improvements in the standards – the better designed the standards, the fewer will be the situations for which they are seen to be inappropriate. The second reason engineering authorities might approve a deviation is that line managers are seeking a relaxation of the standard for commercial reasons and are arguing that the increase in risk is insignificant or at least reasonable in the circumstances. This puts engineering authorities in a difficult position. The problem is that although in any one case it may be true that the deviation does not significantly increase the risk, if such authorisations become the norm, the associated risks will creep imperceptibly upwards. The proposed indicator helps to control this process and strengthens the will of the approver to resist commercial pressures. Furthermore, this is a relatively robust indicator, in the sense that authorisation is a discrete event and hence the number of authorisations is difficult to fudge. (Of course, along with most other indicators, it can be falsified, but that is another issue altogether.) For this reason, it might well be appropriate to make this a measure that matters by specifying targets in performance agreements.

2. Number of authorised safety system defeats (by-passes) that are currently in place (the fewer the better).

Defeat or by-pass refers to a situation in which an acknowledged defence has been disabled for some reason, e.g. for purposes of maintenance. The risk level is ipso facto higher, but a risk assessment in the particular case may deem the risk to be acceptable. However, unless the number of such defeats is monitored and there are incentives to drive the number down, long-term by-passes may become normalised.

3. Safety-critical maintenance backlog.

This indicator is widely accepted[22] as an important indicator of major hazard risk, although it needs to be carefully audited because of its vulnerability to manipulation.[23] (We say more about this in Chapter 7.)

While it may not be possible to develop a complete suite of indicators of barrier health, there are indicators such as those above that can usefully be employed to drive improvements in risk management and may well be sufficiently immune from manipulation to be considered for inclusion in bonuses systems.

Rewarding senior managers for managing risk

It is senior managers who have the greatest capacity to influence major hazard risk. Let us consider therefore how their performance agreements can be used to focus attention on these matters.

It is critical that their performance agreements deal with things that are within their power to influence. From this point of view, the problem with many numerical performance indicators is that it is not immediately obvious how the activities of any given manager may influence them. For operations and engineering managers, some of the site-wide or business-wide performance indicators (such as leaks) may be appropriate measures for inclusion in personal performance agreements, precisely because they have direct influence over these matters. Other managers in senior leadership teams – such as HR managers, finance managers and legal counsel – may feel that they have little or no control over the number of precursor events, or the health of the barriers. An incentive system that highlights these issues therefore has little potential to influence their behaviour. Putting it another way, measures of this nature are not sufficiently targeted to motivate the performance of such managers.

However every major accident inquiry highlights ways in which such managers might have acted to prevent the accident or at least reduce the risk of its occurrence. One issue frequently identified in major accident inquiries is the lack of competence on the part of some or all of those involved. This is an issue for a personnel manager. He or she can influence major hazard risk by ensuring that positions which are defined as safety-critical are staffed by people with the necessary competence. This is a challenging goal. It requires that safety-critical positions be identified, that competency requirements of these positions are defined, and finally that resources are allocated to ensure that people have the necessary competencies. This can even be turned into a numerical indicator – proportion of safety-critical positions filled by people with the necessary competencies. Performance against such a measure is frequently undermined when people are asked to act in higher positions for which they are not qualified, so personnel managers would find themselves taking an active interest in the filling of temporary vacancies. This is just one way in which such an indicator might drive improvements. Including such a KPI in the performance agreements of personnel managers would give them an incentive to play their part in reducing major hazard risk.

A second issue almost invariably identified in major accident investigations is that cost-cutting had undermined the capacity of an organisation to monitor and respond to major hazard risk. This is an issue for finance managers. Top managers often issue challenges to lower level managers to find ways to reduce costs without reducing the level of safety. Lower level managers, whose careers may depend on them delivering on these challenges, do their best, but may find themselves cutting supervision, maintenance and training, all things that increase the risk of a major accident. One possible solution is to make senior finance managers responsible for risk assessing cost cuts that are made in response to such challenges. It would not be enough for a finance manager to seek assurances from line managers that they were making the cuts without jeopardising safety, because line managers will often feel constrained to provide such assurances, no matter what the situation. Rather it will be necessary for finance managers to develop alternative sources of unbiased information, so that they can take responsibility themselves for ensuring that cost cuts are not at the expense of safety. Including such a responsibility in the performance agreements of top finance managers could be expected to provide a powerful counterweight to one of the most commonly identified causes of major accidents.

A third issue identified in many major accident reports is that the organisation had failed to learn from previous incidents and accidents. Interestingly, there is a role here for legal counsel who sit on executive teams. Reports on internal incidents often come to company legal departments to ensure that they are not incriminating, that is, to ensure that if and when released, they do not expose the company to additional legal liability. The lawyers concerned sometimes veto or delete information about organisational causes on the grounds that these are speculative and expose the company to liability. Yet this information about organisational causes may embody the most important lessons for accident prevention – lessons that, if learnt, will protect the company against similar future incidents. The irony, then, is that in seeking to protect the company against litigation arising out of a particular incident, lawyers are increasing the risk of future incidents. The challenge for legal counsel is to balance these two interests in some way, that is, to ensure that the incident reports maximise the learning to be derived from the incident, as well as minimising the risk of litigation. This outcome can be achieved by inserting appropriately worded objectives in the performance agreements of legal counsel. In this way legal counsel, who rarely see themselves as having a role in accident prevention, can be enlisted in the cause.

These are just three examples designed to provide an idea of the potential. The point is that senior managers must have performance agreements that identify safety-relevant activities peculiarly within their control and for which they can sensibly be held personally accountable. It may require some ingenuity to devise appropriate safety-related goals for senior functional managers. Corporate HSE (health, safety and environment) managers can be particularly useful here. Ensuring that appropriate HSE objectives are included in the performance agreements of senior executives is one of the most important functions that a corporate HSE manager can perform. It follows that one of the most important initiatives CEOs can take is to include in the performance agreements of corporate HSE managers a requirement that they identify appropriate HSE goals for inclusion in performance agreements of all other members of the senior management team. Possible items for inclusion in individual performance agreements of senior managers include:

- actively participate in investigations into all major incidents in area of responsibility;

- ensure that data bases of accidents and incidents are analysed for trends and patterns;
- ensure that health and safety are included in the design of new plant and processes;
- ensure that health and safety are included in agreements with suppliers;
- ensure that all immediate subordinates read and discuss a particular book that is relevant to process safety;
- include health and safety in all position descriptions;
- coach and mentor senior managers on health and safety leadership;
- ensure that all positions designated as safety-critical are filled with staff whose competency has been assured;
- scrutinise incentive schemes to ensure that unintended and perverse consequences are minimised;
- evaluate the impact of shift arrangements and overtime on levels of fatigue;
- actively establish that cost cuts are not at the expense of safety.

Once objectives are chosen for the most senior executives, they can be cascaded downwards so that the performance agreements of immediate subordinates encompass matters within their direct control, and so on down the line. In this way, the interests of individuals at all levels can be aligned with those of the organisation.

None of this is to suggest that site-wide or company-wide performance indicators can be dispensed with in the incentive schemes devised for senior executives. If nothing else, these measures serve to remind senior people of overall corporate goals and encourage them to align their behaviour with these goals wherever possible.

Rewarding individual initiatives

The final strategy to be discussed in this chapter is to recognise and reward individuals who take an initiative to reduce major hazard risk. The most important initiative in this context is the reporting of bad news, for the following reason.

Prior to every disaster there are always warning signs – indications that things are amiss. Had these signs been identified earlier, the disaster could have been avoided. Furthermore, people at the grass roots of the organisation are frequently aware of what is happening but do not transmit the bad news upwards, for a variety of reasons.

Mindful leaders are acutely aware of this problem.[24] For them, bad news is good news, because it means that their communication systems are working to move the bad news up the hierarchy to the point where something can be done about it before it is too late. One of us sat in the office of such a leader one day while she was talking on the phone to a lower level manager who had provided her with a report that presented only good news. "But where is the bad news," she said. "I want you to rewrite your report to include the bad news." The organisation in question had a policy of "challenging the green and embracing the red". As noted earlier, this slogan referred in the first instance to traffic light scorecards. But it also had a more metaphorical meaning: question the good news and welcome the bad. She was implementing this slogan in a very effective way.

To encourage the reporting of bad news, organisations must *celebrate* particularly significant reports. There is a famous case in the literature[25] where a seaman on an aircraft carrier thought he might have left a tool on the deck. Foreign objects on a runway are dangerous. Accordingly, the seaman reported the loss of the tool to the commanding officer of the carrier. There were aircraft aloft at the time that had to be diverted to shore base. The tool was found and the aircraft brought back on board. The whole episode involved a substantial disruption of the activities of the aircraft carrier. The next day the commander summoned the whole crew to the deck and held a ceremony in which he congratulated the seaman for having made the report.

This kind of recognition can also involve financial rewards. The mindful leader referred to earlier had introduced an incentive system to encourage the reporting of bad news. She had instituted an award, named after a man in her organisation who had saved someone's life by his alertness to a process safety hazard. The award had various levels, the highest being diamond, which was worth $1000. The day of our visit she made a diamond award to an operator who had recognised that some alarm levels had been changed on a rotary compressor without a proper management of change procedure. He had written an email about this to his manager who in turn had passed it up the line. The senior manager we were visiting had made more than a hundred awards for this kind of reporting in a period of less than 12 months.

The incidents just recounted have two features in common. First, it took courage to report. In the aircraft carrier case, the seaman was reporting his own "carelessness", while in the compressor case the email writer was implicitly reporting on a failure by one or more of his superiors, which could easily have made life difficult for him. The second common feature is that both incident reports were oriented to the prevention of a major accident, not a personal safety issue. This is arguably the most helpful kind of reporting.

The preceding examples tap two different motivations: financial and psychological (public recognition). Presumably the best of both worlds can be achieved by combining the two, that is, providing financial incentives that are backed up by public recognition.

Notice also that because such rewards are at the discretion of senior managers, there is less likelihood of the unintended consequences that can occur when rewards are provided for achieving numerical targets. Where incentive systems reward the *quantity* of reports, not their *quality*, people respond by making reports that may be of little or no value in promoting safety.

The strategy of recognising and rewarding outstanding instances can be expected to generate a culture in which this kind of reporting becomes the norm. Furthermore, if the reports that are acknowledged in this way relate to major hazard risk, not personal safety, people will orient their reporting accordingly.

Finally, we note that reporting is not the only activity that reduces risk. It is also open to organisations to acknowledge and reward any and all outstanding instances of major hazard risk reduction. Again, because such acknowledgement is at the discretion of senior managers it is less likely to result in perverse or unintended consequences.

Conclusion

This chapter has canvassed four different ways in which incentives can be used to focus attention on major hazard risk. The first, suggested by the finance industry, is to defer the payment of bonuses for some years, so that employees, and particularly the most senior executives, have a personal interest in the long-term fortunes of the business. However this is an indirect and somewhat uncertain way to promote a focus on catastrophic risk in the here and now. As we shall see in Chapter 4, whether such schemes can have the intended effect depends critically on scheme details.

The second strategy is to develop indicators of how well catastrophic risk is being managed in the here and now and to reward people based on these indicators. The problem here is to find indicators that are true measures of the relevant risk. In particular each major hazard industry and industry segment needs to identify relevant precursor events from which valid performance indicators can be constructed. Indicators that measure the health of particular barriers can also be developed and aggregated to give a picture of the state of the major hazard risk management system. These indicators are all subject to the problem of unintended effects which may undermine their value entirely unless this problem is explicitly recognised and carefully managed.

Third, the activities of most, if not all, senior managers are directly relevant to major hazard risk. The particular contribution that each can make to major hazard risk reduction can be identified and included in performance agreements. This is a relatively direct way of influencing risk.

Fourth, people can be acknowledged and rewarded when they take initiatives in relation to major hazard risk, in particular by reporting matters that could potentially end in disaster. This is one of the best ways to create a culture focused on major hazard risk. It is also the strategy that is least susceptible to perverse consequences.

Notes

1   Steve Greenhouse, *New York Times*, 30 October 2009.
2   *Financial Times*, 19 June 2007.
3   *Wall Street Journal*, 12 March 2012.
4   http://www.lse.co.uk/ShareChart.asp?sharechart=BP.&share=bp.

5 Some organisations provide loans that can be used to buy shares.
6 Bebchuk and Fried, 2010, p. 6.
7 Yeh, 2010, p. 104.
8 See references in Bebchuk and Fried, 2010, p. 25.
9 Bebchuk and Fried, 2010, p. 9.
10 Yeh, 2010, p. 100.
11 Hopkins, 2008, p. 73.
12 Adapted from Reason, 1997.
13 API 754 speaks of loss of *primary* containment (LOPC), but we shall ignore the distinction here.
14 This is discussed further in Chapter 7.
15 Hopkins, 2009, Chapter 3.
16 See also the discussion of "vent gas" in Hopkins, 2012, p. 89.
17 Diener, 1999, p. 1.
18 For a more detailed discussion see Hopkins, 2012, pp. 89–92.
19 Strictly speaking, mass.
20 Hopkins, 2009.
21 API 754, p. 13.
22 In particular by the UK Oil and Gas Industry. See B. Lauder, "Major Hazard (Asset Integrity) Key Performance Indicators in Use in the UK Offshore Oil and Gas Industry". Paper presented to CSB Meeting, Houston, 23 and 24 July 2012.
23 For an example of such vulnerability, see Hayes, 2013, pp. 100–101.
24 Hopkins, 2007, Chapter 9.
25 Weick et al., 1993.

Chapter 4
Long-Term Incentives

The preceding chapters serve as the introduction to the empirical study we described in Chapter 1. The aim of the study is: to identify in greater detail the incentive schemes that companies in hazardous industries are using; to analyse the ability of these schemes to influence major hazard risk; to identify best practice or lessons with respect to the management of catastrophic risk; and finally, to investigate how people actually respond to incentives. We aim both to learn from what industry is doing and to offer constructive criticism. To begin with, therefore, we must outline the architecture of the remuneration systems of the companies we studied.

The architecture of remuneration

Bonus payments vary with a person's position in the company hierarchy. In nearly all cases the CEOs received bonuses that far exceeded their fixed pay. The further down the hierarchy the individual was, the smaller was the bonus paid, and workers on wages were sometimes ineligible for any bonus at all.

With possibly one exception, all companies operated two distinct bonus schemes: a short-term, incentive program (STIP) and a long-term incentive program (LTIP). The STIP payments aimed to promote behaviour consistent with the company's interests in relation to a range of issues, including production, profit, environmental compliance and safety. They were available to all employees in management positions and sometimes even to front-line workers. STIPs were paid annually, at the end of the year in which they were earned, except for the most senior executives, some of whose STIP payment was deferred for a period of years.

On the other hand, long-term bonuses were generally paid only to the CEO and top managers. They were paid exclusively on the basis of financial performance (with one or two exceptions to be noted later). For publicly traded companies, the payment consisted of a certain number of shares, which were handed over (vested) some years after they were notionally earned. (This will be explained in more detail shortly.) Interestingly, there were no cases in our sample of bonuses paid as share options, as described in Chapter 3.[1] We speculate later on the reasons for this.

For most of the companies in our sample, the main purpose of their incentive schemes is to align the interests of decision makers with those of the shareholders. The word "alignment" was used in nearly every case, and some companies described this aim as a "philosophy". Let us explore this idea for a moment. The literature on business firms, particularly publicly listed companies, speaks of the "principal/agent problem", the principal being the owner (or owners, for example, shareholders), and the agent being the CEO (and perhaps senior executives) hired to act in the interests of the owner. The "problem" is that, left to themselves, agents may act in their own interest, enhancing their own wealth, power and prestige, possibly at the expense of owners. Tying CEO remuneration to company financial performance is a way of solving the principal/agent problem.

This solution has important implications. It encourages CEOs to consider the interests of workers or customers *only in so far as* it is in the interests of shareholders that they do so. Although corporate law requires that CEOs serve the interests of shareholders exclusively, CEOs have sometimes strayed from this principle. A famous example is Henry Ford who, as CEO of the company he founded, chose to pay workers substantially more than the market required, as well as selling his cars to customers as cheaply as he could. His philosophy was that the company should make only a "reasonable" profit, and that "business is a service, not a bonanza". His policies infuriated some shareholders who took him to court for failure to put their interests first. The disgruntled shareholders won and Ford was forced to accept that he had a legal duty to put shareholder interests above all others and that he could not run the company he founded as a benevolent institution.[2] By implication CEOs must not devote more resources to safety and environmental matters than is in the interests of shareholders. The remuneration systems under consideration are implicitly designed to achieve this outcome.

One way in which governments have tried to make safety and environmental matters a higher priority is to enact safety and environmental laws and to prosecute violators. Where companies are the target of prosecution, the fines involved are often too small to make an impact on corporate profit and are therefore largely ineffective as a deterrent. However when CEOs and other very senior managers are the target, the deterrent effect is more powerful. The prosecution of individuals can be seen in this context as an attempt to split the alignment of interest between the top executives and shareholders and to provide the former with an incentive to behave in ways that may not be in the interests of the latter.

But to return to the issue of bonus payments, in order to understand their full significance we need some idea of their relative size. The proportion of total remuneration coming from bonuses varied across our sample and varied with position in the company. To simplify matters, let us consider the situation of the CEO only. In general we found that total bonus payments available to the CEOs of publicly listed companies were several times greater than the fixed salaries they were paid! In other words, a very large proportion of their remuneration was "at risk", to use the jargon, dependent on CEOs doing whatever it was they were incentivised to do. This is one of our most striking findings.

It is important here to break the bonus into its short-term and long-term components. The maximum short-term incentive payable was often equal to at least the base salary and in one case, six times the base salary! Long-term incentives were larger, with maximum bonuses often three or four times the base salary. Again, in one case the maximum long-term bonus payable was six times the base salary! On the whole, then, long-term incentives were substantially larger than short-term incentives.

This means that long-term incentives should figure prominently in any discussion of the impact of bonuses on safety. This is especially true given that long-term bonuses are tied exclusively to financial performance, which, on the face of it, would predispose CEOs to put profit ahead of safety. Yet most discussions of the impact of bonuses on safety focus on short-term bonus systems, perhaps because many more employees are eligible for short-term bonuses. Long-term incentive payments are thus the elephant in the room. Unless the elephant is acknowledged, discussions about the impact of bonuses on safety have an inevitable air of unreality about them. Let us therefore confront the elephant.

Long-term incentive payments

To understand the potential effect of LTIPs we need to understand in more detail how they work. The single most common scheme was as follows. Each year, a number of what are sometimes called "performance share awards" are provisionally made to the CEO and some key executives who are judged to have the capacity to influence business outcomes. The number of shares depends on the level of the individual in the hierarchy and on the business performance that year. This in itself provides some incentive to maximise business performance in the year of the award, but that is not the main incentive effect of these shares. The shares are only provisionally awarded and "vested", that is, are actually handed over, some years later, typically three, if certain conditions are met. This is where the real incentive to maximise financial performance lies. The main condition relates to – total shareholder return (TSR). But it is not simply the TSR that is relevant, rather the *relative* TSR, that is, the return relative to some comparator group. That group may be a group of similar companies in the same industry or it may be less industry specific, such as the top 100 companies on the stock exchange. The total shareholder return of the median company in the comparator group is identified and becomes an index against which the TSR of the company concerned is compared. The comparison works something like this. If, in the year the shares are due to vest, the TSR is less than the median or index value, none of the performance shares vest, that is, the CEO and senior executives get no benefit at all from the performance shares awarded three years earlier. If the TSR achieves some specified value above the index, for example if it is located at or above the 75th percentile, then all the shares vest. A result that sits between the 50th and 75th percentile attracts a percentage of the available shares, starting at say 50% for performance at the 50th percentile and rising to 100% at the 75th percentile, on a sliding scale. This situation is represented in Figure 4.1.

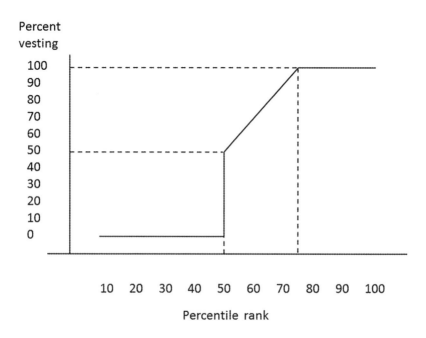

Figure 4.1   LTIP vesting – paradigm case

We can summarise this in general terms as follows. If company performance is below the middle of the comparator group, senior executives get nothing. If it is substantially above the middle, they get all their performance shares. And if the performance comes in somewhere between these two possibilities they get some proportion of the award, on a sliding scale.

A variation on this is based on percentage above the average TSR of the comparator group, rather than the percentile ranking. Such a scheme is presented in Figure 4.2. In this case, for the shares to vest fully, the TSR must exceed the average by 6% annually.

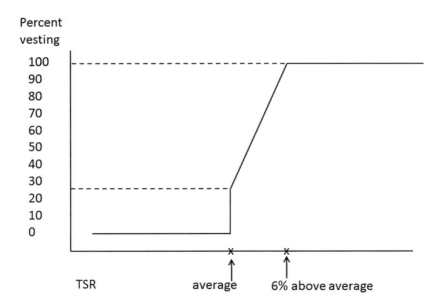

Figure 4.2   LTIP vesting – alternative framework

In order to understand the significance of these schemes let us consider two limiting cases, presented in Figure 4.3.

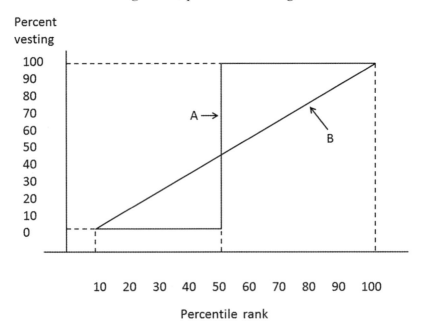

Figure 4.3   Limiting cases for LTIP vesting

We did not encounter either of these limits in our research, but it is useful to consider these hypothetical cases. The scheme depicted by line A provides no bonus till the performance reaches the middle of the comparator group, after which the bonus jumps to 100%. This scheme puts extreme pressure on the CEO to ensure that the company's performance reaches the 50th percentile, but no incentive to push for higher shareholder returns, once the 50th percentile has been reached. It is perhaps because of this latter feature that this scheme is not favoured by the designers of incentive schemes. At the other limit, line B represents a bonus that increases steadily from 0% to 100%. This means that at whatever level the company is performing, there is always an incentive to perform better, but there is no particular pressure to get across the 50% threshold. Every incremental improvement reaps a reward for the CEO, but it is a relatively small reward. Such a scheme provides a continuous but relatively gentle pressure to improve performance.

The schemes represented in Figures 4.1 and 4.2, lay between these limits. Because they provide nothing till the 50% threshold is crossed, they put extreme pressure on CEO's to ensure that their companies get to at least that threshold. Moreover, because there is still a sloping line above that threshold, they provide an additional incentive to push the performance above average. Every increment above the 50% threshold yields additional bonus, until at some point the CEO receives the bonus in full (100%). These schemes were clearly designed to maximise these incentive effects.

One company we studied had fine-tuned the "slope", presumably to try to maximise the incentive effect. It had previously been operating on the basis of line one in Figure 4.4 but had recently changed to the scheme described by line two. The differences are instructive. The step up at 50% threshold has been increased (from a third to a half), providing an even greater incentive to cross the threshold, and the slope above that point has been steepened, increasing the pressure to improve performance, up until the 75th percentile. There is no incentive to improve performance beyond that point. The scheme designers have chosen to "use up" all the incentive effect to encourage CEOs

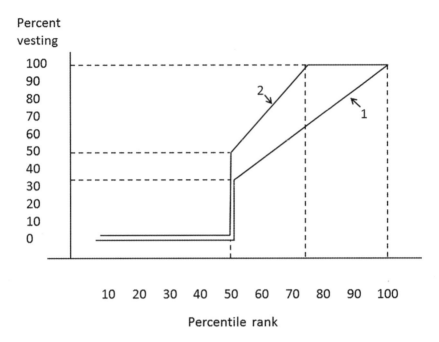

Figure 4.4   LTIP vesting – before and after fine tuning

to get to the 75th percentile. Putting it another way, they will be fully satisfied if the CEO achieves top quartile performance.

These schemes, then, are not gentle in their effects. A small difference in TSR can make all the difference between zero vesting and 100% vesting. The result is that senior executives are under enormous pressure to ensure that their companies perform significantly better than their comparators. This is presumably the intention of those who design these schemes. This is a high-stakes game with no consolation prizes for the losers.

It is noteworthy that none of the companies in our sample was using share options as described in Chapter 3, although some had previously used option schemes. It will be remembered that option schemes require only that the share price be higher at the time of vesting than it was at the time of award, in order for the recipient to make a profit, and that the higher the share price at the time of vesting, the greater the profit. This is a somewhat gentler pressure than that described above, without the same competitive edge of share awards that are subject to *relative* total shareholder returns. We can assume that this is one of the reasons that share option strategies were not in use in the companies we studied – they did not put enough pressure on CEOs to maximise financial performance. Another reason is that share prices are not only influenced by the efforts of the CEO; they are subject to the vagaries of the market. If resource prices collapse, the share price of all relevant resource companies will collapse. To penalise CEOs in these circumstances is counterproductive. Using *relative* TSR as the basis for bonus means that high-performing CEOs can still receive a long-term bonus even though share prices have fallen.

Consider now the impact of these long-term schemes on safety and environmental performance. Most safety and environmental issues do not have the potential to affect shareholder return. The logic of performance share awards therefore encourages senior executives to minimise spending on safety and environmental issues in order to maximise shareholder return.

However the situation is not so obvious for catastrophic events, which *do* have the capacity to impact on share prices. Some of our interviewees suggested that bonuses based on financial returns provided an incentive for senior executives to pay attention to catastrophic risk, because of the financial cost of major accident events. In particular, a catastrophic event will probably ensure that shareholder return drops below average. The long-term bonus will therefore drop to zero, possibly for some years. That, in itself, might be expected to provide an incentive to manage catastrophic risk effectively. On the other hand, if the company is reorganised to behave more conservatively with respect to catastrophic risk, the financial cost of this reorganisation might well reduce shareholder return to below average, and the long-term bonus again drops to zero. Interestingly, BP explicitly acknowledges this problem in its 2012 Annual Report.

> Performance related to restoring value (after the Deepwater Horizon disaster) was somewhat mixed, in part reflecting the priority throughout the company's business of continuing to embed safe and effective operations ....[3]

In other words, the priority BP had given to improving the management of catastrophic risk had had a detrimental effect on the performance of various business indicators including total shareholder return. There is no doubt that the shock of the

Deepwater Horizon accident was sufficient to ensure that BP executives focused on catastrophic risk, despite any potentially negative effects on their long-term bonuses. In general, however, tying long-term bonuses to relative shareholder return provides little incentive to focus on catastrophic risk. On the contrary, the likelihood of suffering a catastrophic event is low, while the likelihood of falling below the comparator group average – and suffering total bonus wipe-out – is obviously much higher. In principle this encourages CEOs to try to maximise TSR even at the expense of catastrophic risk management. In this way the long-term bonus schemes under discussion seriously compromise catastrophic risk management. We shall return to this issue in our concluding chapter.

Recall, now, that the long-term incentive system in particular is intended to align the interests of CEOs with those of shareholders. Let us pause therefore to consider the shareholder's situation. A shareholder return that is a little below the average may still be a healthy return. Moreover, whether the return is a little above or a little below the average does not greatly affect the shareholder's circumstances. It follows that bonus designers, who have made the achievement of at least average TSR such a critical issue for CEOs, have over-reached themselves in seeking to align the interests of principle and agent. They have ended up creating a system in which the two sets of interests are systematically *out* of alignment.

This is not just a theoretical point. This lack of alignment creates different interests in relation to the management of catastrophic risk. As explained above, the performance bonuses under discussion encourage CEOs to take greater risks in order to end up in the top half of the distribution. These may be commercial risks, but they may also be safety and environmental risks. Shareholders do not have this motivation to take risks. On the contrary, a major accident can have a dramatic impact on share prices and shareholders therefore have an interest in their company taking a relatively conservative approach to the management of catastrophic risk, even if that means slightly lower returns. Some of the largest investors are superannuation or pension funds, investing the contributions of countless wage and salary earners. Moreover pension fund contributors frequently opt for a *balanced* investment strategy, rather than a *growth* strategy that promises higher returns but at greater risk. The performance share award schemes under discussion are quite contrary to the interests of such investors.

Variations and qualifications

The preceding discussion has ignored numerous variations in order to focus on the main features of the LTIPs in our sample. Let us sketch some of this variability.

Other business measures

For some companies there were additional business measures apart from TSR that were included. However TSR remained the most important and all the additional measures seemed to include the sudden death cut-off if performance fell below the comparator group average. There was thus no letup in the competitive pressure generated by such schemes.

An exceptionally small comparator group

For one company, the chosen comparator group was so small, consisting of only five companies, that it made little sense to talk about percentiles or averages. Instead it simply rank ordered the comparator group, on the basis of performance over the previous three years. If the company under discussion was ranked first, top execs received twice the number of the shares nominally awarded three years earlier, if second, 1.5 times the original award, if third, 0.8 of the original award and if 4th or 5th, they received nothing. This is another version of the sudden death cut-off described earlier: if the company falls below the median ranking, the bonus is wiped out. The most significant difference in this case is that once the threshold median value is achieved the bonus rises with every step up in ranking and only achieves its maximum value if the company ranks first in its comparator group. The CEO is therefore incentivised not just to do better than most of his competitors but to be the best.

Safety as part of the LTIP?

Some companies told us at interview that safety was incorporated into their long-term bonus arrangements. However this was in the context of evaluating overall business performance for the purpose of deciding how many bonus shares were to be initially awarded. Safety played no role in determining the extent to which those shares would finally vest – that depended purely on financial performance. In other words, even where safety was nominally relevant, it in no way countered the pressure that CEO's experienced to ensure that the company's financial performance was above the comparator average.

Vesting period

The most common vesting period was three years. However one company used five years. One of our interviewees suggested that the vesting period should be up to ten years, so that the payments remained at risk for a much longer period. His reasoning was that the damaging effects of cost-cutting decisions might take up to ten years to manifest themselves. A ten-year vesting period would give senior executives a greater incentive to manage catastrophic risk more effectively, he thought. We return to this issue in the concluding chapters.

Other share schemes

Apart from long-term incentive plans described above, several companies operated other schemes aimed at encouraging employees to build up share holdings that would give them a long-term interest in the company. One such scheme allowed eligible employees to buy shares from the company out of pre-tax earnings. In addition, CEOs were often required to build up and retain substantial shareholdings in the company. These various schemes reinforced the interest of those concerned in the financial well-being of the company, although they could have only very limited incentive effect since most beneficiaries could have no discernible impact on company fortunes.

Deferment of a proportion of the short-term bonus

Some companies chose to award part of the STIP in the form of deferred shares. One model was that half the short-term bonus would be deferred in this way. Another was that, where the total bonus was greater than the base salary, the amount in excess of the base salary would be deferred. These deferred shares vested after two or three years. Their final value thus depended on share price movements during the two or three years vesting period. The delay provided beneficiaries with an incentive to maximise shareholder return during the vesting period. Interestingly, therefore, although the original STIP award may have provided some incentive to improve safety, this was subtly undermined by the fact that the way to maximise the value of the award, once made, was to maximise shareholder return. In other words, the mechanism of deferment turned part of the short-term bonus into a longer-term bonus dependent on financial performance. The net effect was to accentuate the imbalance between the short-term incentives in which safety was taken into account, and long-term incentives which were for all practical purposes exclusively dependent on financial performance. Where short-term incentive payments are deferred in this way, the shadow cast by the elephant in the room is even longer.

Staff retention

Although long-term incentive programs are primarily designed to maximise shareholder return, that is not their only purpose. For many of our companies, a secondary purpose was to retain employees. At any one time, each eligible employee has perhaps three years' worth of bonus shares potentially owed to them. If they leave without good reason (and moving to another firm would not be judged a good reason), they stand to lose these entitlements. Such an arrangement encourages loyalty to the corporation and discourages executives from using their market situation to maximise remuneration. For the privately owned company discussed below whose long-term incentive scheme was available to all employees, this staff retention function was seen as perhaps the primary purpose of the scheme.

An outlier

One of the companies we studied had a very different long-term bonus structure. There were several interesting differences. First, it was a private company, owned by two larger companies, and therefore not in a position to offer bonus shares. Its LTI bonus was paid in cash. The cash amount was determined by business performance, mainly financial performance, in the year of the award. The award amount was also marginally influenced by a process safety indicator – loss of containment. Payments were paid over three years in equal amounts. But because they were cash payments, not shares, they were not affected by business performance during that three-year period. Consequently they could have no incentive effect during the period between when they were awarded and when they were received.

A second difference is that, with one exception to be discussed below, the scheme applied to all employees, not just those at the top of the hierarchy whose behaviour had the potential to make a discernible difference to the company fortunes. For most people, therefore, these awards provided no direct financial incentives to behave in any particular way.

Third, the amount of the bonus was about 10% of salary, from top to bottom of the company, with the exact percentage depending on business performance. In short, relatively little was at stake.

The one exception noted above was the CEO. He was a secondee from one of the owning companies. His long-term bonus was paid in shares of the parent company and was linked to both the performance of the parent company and his individual performance. We were not in a position to get more details of his remuneration scheme and are therefore not in a position to pinpoint its possible incentive effects.

The company's LTIP seemed to operate more as a profit sharing arrangement than as an incentive scheme. Insofar as that is an accurate characterisation, the scheme did not serve the interests of the owners, in fact was contrary to the interests of owners, and might not have survived had this been a publicly listed company.

Conclusion

Long-term incentive plans tend to be passed over in most discussions of the impact of bonuses on safety. They shouldn't be. Long-term bonuses are the biggest single component of CEO remuneration and their impact on safety needs to be carefully considered. The fact is that long-term bonuses are exclusively dependent on financial performance; safety performance is essentially irrelevant. There is one qualification to this conclusion. Where an accident event is of sufficient magnitude, it may affect shareholder returns. On the face of it therefore, long-term bonuses provide CEOs with an incentive to invest in measures that reduce the risk of major accidents. However, long-term incentive schemes are designed to put CEOs under enormous pressure to achieve better than average financial performance, in relation to some relevant comparator group, every year. If they fail to get across the threshold their bonus is wiped out – totally. A CEO therefore faces a difficult financial risk management decision. We put it in the first person below in order to dramatise it.

> Should I invest the additional millions, or perhaps billions to reduce the risk of a major accident, or not? An accident that impacts on shareholder returns will wipe out any chance I have of receiving my long-term bonus. On the other hand the probability of such an accident occurring on my watch is low. Moreover, if I invest the funds necessary to reduce that risk further, this will probably reduce total shareholder return, increasing the risk that the company will drop into the lower half of the distribution, again wiping out my bonus. This latter possibility is a far more real and present threat to my long-term bonus than is the risk of an accident sufficiently serious to impact shareholder returns. Faced with these alternatives, I therefore choose to maximise *current* TSR, regardless of the long-term safety consequences.

Such is the logic of the long-term incentive schemes that many Boards have designed for their CEOs.

Notes

1 One company provided loans to buy shares. This had some of the features of a share option scheme.
2 Bakan, 2004, pp. 34–37.
3 BP annual report for 2012, p. 132.

Chapter 5
Annual Bonuses

Annual bonuses (short-term incentives) are awarded to people at many levels in the hierarchy, not just to the CEO and direct reports. The stated aim of these bonuses is to align the interests of as many employees as possible with corporate goals. For this reason they are the subject of constant debate and adjustment. This is one of our most striking findings: the fact that short-term incentive plans were constantly changing, as companies sought to refine them. What we report on here is therefore a snapshot in time.

Unlike long-term incentives, annual bonuses *were* significantly affected by safety. Our ultimate goal here is to understand in what way and with what potential effect. But in order to do this we need first to describe the architecture of these short-term incentive plans. That is the purpose of the first part of this chapter.

Nearly all companies subscribed to the theory that annual bonuses should include two kinds of measures – measures of individual performance and measures of group performance. Rewarding individual performance has obvious incentive effects. But the case for including *group* performance measures is more problematic. We deal first with this.

Group performance

The first issue in determining group performance is to decide on the group. At one end of the spectrum the group could be a face-to-face work group or team; at the other, it could be the whole global corporation. Global corporations often consist of various constituent businesses, operating in different parts of the world and/or engaged in very different kinds of business activity.

These constituent businesses occupy an intermediate position on the group spectrum, between team and global corporation. Interestingly, the global corporation is often called the Group, with a capital G, to distinguish it from other kinds of group. We shall follow this capitalisation convention carefully to minimise confusion.

The group measures contained in annual bonuses – such as profit and production figures and safety statistics – must relate to one or other of these groups. The question is: which? None of the annual bonuses we encountered was based on teams, so we pass over this possibility here. (An interesting example of bonuses paid to drilling teams was discussed in the addendum to Chapter 2, but these bonuses were paid for specific drilling assignments, not for annual performance.) Most companies used either business units or the global corporation as the basis of their annual bonus system. Some used a mixture of global and business unit measures. One company used three groups: the global corporation, the business unit and the local asset. We concentrate here on companies that used either the global corporation or the business unit.

The argument in favour of using measures of Group or global performance is that this encourages everyone to identify with the interests of the global corporation, not just with the interests of the particular business unit to which they belong. In particular, it encourages heads of business units to cooperate, rather than operating their businesses as "silos", without regard for each other. Moreover, if one business unit is less profitable than another through circumstances that are quite beyond the control of its management, senior corporate managers will be more willing to accept assignments in that business unit if they know that their remuneration is tied to the performance of the corporation as a whole, rather than to that of the less profitable constituent business. All this was put to us by our interviewees in justification of using Group or global measures.

The problem with using Group performance measures, however, is that the great bulk of employees, including most senior managers, feel they have no ability to have a perceptible influence on outcomes. The word "perceptible" is important here. It is true that global performance is the outcome of the individual efforts of myriads of people, but if not one of these people has any sense of a connection between his or her efforts and the global result, then the global result cannot act as a motivator. The term used by our respondents to describe this situation was "line of sight". If managers have no line of sight to outcome measures, so the argument went, there was no point in including such measures in incentive programs. From this point of view, said some of our respondents, it was preferable to use business unit measures of performance, rather than global. Unfortunately, though, this does not solve the problem. The constituent business units of global corporations are generally too large for most staff to have any sense that they can influence outcomes. Hence, treating the business unit as the relevant group for the purposes of performance measurement is hardly any better than using Group level measures.

This is a much debated issue in many of the organisations we studied. Most were using global corporate measures, but some had only recently adopted this approach. However, if none of the group measures used in annual bonuses has the capacity to motivate behaviour for most staff, the debate is pointless. To foreshadow our conclusion, this is an argument for abandoning the group component of annual bonuses altogether (except for top management – see below).

Once the group and the group measures have been identified, the next step is to convert these measures to a single indicator of group performance. Many companies had complex ways of doing this and we shall skim over this complexity here. Suffice it to say, they set target levels in relation to each of the relevant measures (such as profit, production and injury rates) and calculated scores for each measure in terms of whether it was above, below, or on target. These scores were weighted according to the importance given to each of the constituent measures, and a weighted average calculated. The system was often designed so that a performance that is overall on target yields a final score of 1; an above-target performance yields a number a fraction above 1, while a below-target performance yields a number a fraction below 1.[1]

Table 5.1   BP's annual bonus scorecard results for 2012[2]

Outcomes

2012 annual bonus outcomes

Measures	Weight	Outcomes relative to plan		
		Threshold	Target	Max
Safety and operational risk management	**30.0%**			
Loss of primary containment	10.0%			■
Process safety tier 1 events	10.0%			■
Recordable injury frequency	10.0%	■		
Rebuilding trust	**20.0%**			
External reputation	10.0%		■	
Internal morale and alignment	10.0%		■	
Value	**50.0%**			
Operating cash flow	11.7%	■		
Underlying replacement cost profit	11.7%	■		
Total cash costs	11.7%	■		
Gearing	3.0%		■	
Divestments	3.0%	■		
Upstream unplanned deferrals	3.0%		■	
Upstream major project delivery	3.0%		■	
Downstream net income per barrel	3.0%			■
Overall outcome			■	

Given that this book is inspired by BP's experience, it is worth noting that BP makes use of this approach. It presents the results of this group evaluation process in 2012 in its annual report, reproduced here in Table 5.1 opposite.

On the left of the panel are the performance indicators, in the middle, the weights, and on the right are the outcomes relative to set the targets. According to the text in the report, calculations yielded a weighted average of .97 or 97% of target, represented by the rectangle on the bottom line.

It has to be said however that none of the systems we examined generated this final number in an automatic way – there was always an element of discretion or judgment involved. Even BP relies ultimately on a subjective assessment, despite the elaborate quantitative approach presented in Table 5.1. Here are some observations from a recent BP remuneration committee report.

> At the end of each year, performance is assessed relative to the measures and targets established at the start of the year, adjusted for any material changes in the market environment. Assessment includes both quantitative and qualitative views as well as input from the other committees on relevant aspects. The committee consider that this informed judgement is important to establishing a fair overall assessment.[3]

One company provided us with a revealing glimpse of how this worked in its particular case. In arriving at a final group result the remuneration committee took note of:

- Two fatalities;
- TRCF "well above our target";
- "disappointing production performance";
- "good progress" bringing new projects into production;
- a highly successful drilling program;
- a successful portfolio rationalisation program;
- financial results "slightly ahead of budget".

Some of these elements were associated with numerical targets but others were not, and overall there was no way that a final evaluation could just be read off from the data. Accordingly, the committee proceeded as follows. First, it examined all the evidence, both quantitative and qualitative. Then, guided by the weightings, it made a subjective judgement that, overall, the company had performed marginally above target. Finally, it chose a group multiplier of 1.09 to best reflect this judgment. (It is noteworthy that the poor safety result – two deaths and an injury rate well above target – did not pull the final figure down below one. We shall say more about the relativity of safety in bonus systems shortly.)

To summarise, although many companies employed quantitative measures as far as possible to arrive at a number representing the company's performance, the final evaluation necessarily involved subjective judgements. Sometimes what seemed to happen was that remuneration committees would examine all relevant data, form a view and then choose a single number to represent this view.

Individual performance

We come now to measures of individual performance. Here we need to make a distinction between individuals who are senior enough to influence the performance of some relevant business unit, and most other individuals, who are not. For the sake of brevity we call the latter "ordinary employees".

There is no way that an ordinary employee's contribution to the group profit or the group accident rate can be established quantitatively. Accordingly, senior managers must translate group objectives into specific tasks or projects to be carried out by ordinary employees, in support of group objectives. These specific objectives are formulated as performance agreements between the individual concerned and his or her supervisor. Such agreements often amount to a description of the work to be carried out by the individual during the year. At least once a year the individual meets with his or her supervisor who carries out a performance evaluation against the criteria specified in the agreement. This is necessarily a subjective, rough-and-ready process. Performance is initially judged as falling into one of four categories (sometimes five): such as the following:

Exceptional;
Exceeds expectations;
Meets expectations;
Below expectations.[4]

It is not however left to the supervisor to decide alone on the final evaluation. Individuals are grouped into homogenous comparison groups and the expectation is that the distribution within any one such group will be approximately a bell curve, with say 70% categorised as "meeting expectations". Generally, the supervisors of all the individuals in the comparison group meet to thrash out the final ranking and supervisors have to accept that some of the individuals they wanted to rank as "above expectations" might need to be downgraded to "meets expectations" in order to satisfy the requirements of the distribution.

These rankings must then be converted into numbers for the purposes of bonus calculations. How this was done was not always clear to us, but some of the most explicit schemes functioned as follows. In each comparison group, people are rank ordered, and each is provided with a number that corresponds to his or her rank order. These numbers may vary from zero to perhaps 2, but they must average about 1.0. The purpose of this more finely tuned ranking will be explained shortly.

Needless to say, the whole ranking process is unpleasant for supervisors and generates a high level of resentment among those being ranked. We shall demonstrate this in more detail in Chapter 6.

Consider now the situation of managers who are senior enough to be able to influence the performance of some business unit, in particular, the business unit manager. At this point an interesting cross-over between group and individual performance occurs, precisely because the individual can be held accountable for the group performance. Typically what happens is that a Group-level senior executive will negotiate a performance agreement for the business unit manager that includes numerical targets in relation to production, profit, accident rates, and so on. The agreement may also include qualitative goals such as the completion of a particular project or the implementation of a new system. The latter are truly personal objectives, but the numerical targets are really business unit performance targets for which the business unit leader will be held to account. Business unit leaders in turn develop similar agreements with any direct reports who themselves can reasonably be held accountable for the performance of subgroups. In this way, numerical targets may cascade downwards, although the point is quickly reached where numbers are no longer meaningful and business objectives must be stated in qualitative terms. Simply put, where the performance of a group can be expressed in numerical terms, it is reasonable to hold the group's manager responsible for those numbers; where group performance cannot be quantified in this way, this type of accountability is not possible. But even though senior managers may have numerical targets, these cannot generate performance evaluations in any automatic or mechanical way. The most senior managers must be ranked, just as they rank their subordinates, on a subjective four- or five-point scale.

Combining individual and group

Individual and group performance must be combined in some way for the purposes of bonus determination. We encountered two strategies for doing this – additive and multiplicative. Consider, first, the most common strategy in our sample – the multiplicative. We described above how both individual and group performance were reduced to single numbers. Multiplying these two numbers together yields single "multiplier" to be used in the final determination of bonuses. To give a concrete example based on the scheme described above, if the group performance is on target, the single number will be 1, and if the individual performance is judged to be in accordance with expectations, the single number will be close to 1 (depending on just where the person has been ranked relative to peers). Multiplying these numbers together yields a single multiplier of 1 (or thereabouts).

What, then, does the multiplier multiply? Many companies define a target bonus for each individual for this purpose. Where the multiplier is 1, as described above, the individual receives his or her target bonus. Where the multiplier is greater than or less than 1, the bonus received will be greater than or less than the target bonus, respectively.

The target bonus is often expressed as percentage of salary, and the more senior the employee, the greater the percentage. In this way the target bonus becomes an increasingly important component of remuneration as one moves up the hierarchy. But this is just a start; many other factors go into determining the target bonus for each category of employee. For very senior managers, and in particular the CEO, a major consideration is that the percentage of at-risk remuneration should be considerably greater than the fixed pay, so as to maintain alignment with shareholder interests. Another consideration is the labour market: some types of employee are in high demand and their total remuneration must be kept comparable with whatever competitor companies are offering. Providing larger target bonuses for such people ensures that where company performance is on target and their own performance meets expectations, their total remuneration will be competitive. Interestingly, then, in these circumstances, the target bonus is serving not only as an incentive to work harder, but also as a salary supplement, aimed at retaining valued employees.

Table 5.2 shows how this worked in the case of one company. In particular it shows the multipliers to be used for each combination of group and individual performance. (We did not discover why there are two figures in some boxes.) The multipliers in this figure are not determined quite as described above. Among other things, the individual scores are not as finely tuned, but the scheme is presented in a matrix form that enables us to make certain points.

**Table 5.2   Multipliers used for bonus determination –
a multiplicative scheme**

Organisational Performance	Individual performance			
	Below Expectations	Meets Expectations	Exceeds Expectations	Exceptional
Exceptional	0	1.5	2	2.5
Exceeds Expectations	0	1.25 or 1.5	1.5 or 1.75	2
Meets Expectations	0	1	1.25 or 1.5	1.75
Below Expectations	0	0.5	1	1.5

Notice, first, that if the company performance is on target and the individual meets expectations, the multiplier is 1 and the individual receives the target bonus. Second, if the individual performance is below expectations, the bonus will be zero, no matter what the company performance. Individuals cannot therefore take a free ride and hope to gain a bonus based on company performance. Third, an exceptional company performance coupled with an exceptional individual performance yields a bonus of 2.5 times the target bonus. For senior managers in this company, the target bonus was about 80% of salary. This meant that if both business unit and individual performance was exceptional, such a manager could receive a bonus equal to twice his or her fixed salary. Clearly, in this case the bonus was a very important part of total remuneration. Finally, note that if company profits are down, as represented in the bottom line, bonus payments are automatically reduced, and the majority of people (meets expectations) will get only half their target bonus.

The alternative to all this was an *additive* scheme, which involved weighting each score according to some formula, and then *adding* them together. Various companies adopted this strategy, with weights varying according to the individual's function and position in the hierarchy. For example, one company specified that for the most senior managers, 30% of their bonus would be determined by the individual evaluation and 70% by group measures, while for lower level employees the split was 40%:60%. This differentiation between the most senior managers and others lower in the hierarchy was an implicit recognition that lower level employees could have less influence on group measures than senior executives. Nevertheless for lower level employees the weighting for group measures was still more than 50%. Given that these employees in fact have no perceptible effect on group performance, one wonders why this weighting was not a lot less.

The relative importance of safety – group level measures

Against this background we can now begin to examine in more detail the place of safety in annual bonus schemes. We consider first, group level measures. The issue here is the relative weighting to be accorded to safety or more generally HSE (health, safety and environment) among the various indicators of group performance.

The reader will recall from the weightings in Figure 1.1 that safety and risk management counted for 30% of BP's scorecard in 2012. This included indicators of both personal and process safety, an issue we shall address shortly. Right now what is of interest is the weighting provided to the whole HSE category (30%). None of the companies we studied exceeded this figure. Most were in the vicinity of 20% to 25%, with some as low as 10%. In general, the remaining 75–90% of the scorecard depended on business indicators, such as those contained in the value-creation panel in Figure 1.1. Many of our interviewees endorsed these proportions and told us that their companies had "got it about right". By this they meant that companies were in business to make money and that the primary emphasis therefore had to be on business indicators. On the face of it, however, safety indicators are swamped by business indicators in most company scorecards. Moreover, as we saw earlier, remuneration committees are guided by the quantitative data, but those data do not determine the outcome in any automatic way. This means that it is hard to know whether safety is in fact given the weighting it nominally receives.

Given this tendency for safety indicators to be swamped, companies almost inevitably give priority to profit whenever there is a conflict with safety. This is not to say that companies fail to take safety seriously. It is just that, faced with decisions about whether to spend millions of dollars to reduce safety risks, the scorecard weightings amount to an argument against such expenditure. The Texas City and Gulf of Mexico stories are full of examples of this kind of decision making, that is, decisions to save money by deferring or vetoing safety-related expenditure.[5]

There was one company in our sample that had managed to ensure that the safety component of the bonus was not swamped by indicators of profit and production. They began one recent year with a system much as described above, with HSE contributing 15% to the group scorecard. They decided that this was not giving a strong enough message and considered raising the figure to

30%. However, there was resistance to this idea. They therefore changed strategy altogether. They observed that safety scores would inevitably be swamped when *added* to financial scores and they realised that the problem could be solved by using a *multiplicative* strategy. It is somewhat surprising that while many companies used a multiplicative strategy to combine group and individual scores, only one company recognised that this was a useful way of combining business and safety scores in group scorecards. Here is what it did. First, it developed a single safety measure, based on the total recordable injury rate and the high potential incident frequency rate. Next, it identified a target score for this measure. An on-target score gave a safety multiplier of 1, while above or below target scores gave multipliers of greater than and less than 1 respectively. This safety multiplier was used to modify a single business performance score. A below target safety score therefore reduced the business performance score while an above target safety score raised it. As a result, maximising the value of the business indicator required maximising the value of the safety indicator as well. This was the only company in our sample for which the safety component in the group scorecard was not outweighed and potentially swamped by the commercial indicators. We were told, moreover, that as a result of this change to a multiplicative strategy, people throughout the organisation began taking safety rather more seriously.

The safety indicators in use for group measures

Consider now the safety indicators actually used in our sample of companies to determine group safety performance. By far the most important indicator was total recordable case frequency (TRCF),[6] or some variant thereof. In most companies this was well understood to be a measure of personal safety, not process safety, and several companies were also trying to take process safety into account, as will be discussed below. But even for these companies the TRCF was the single most important indicator.

There was, however, a recognition that, even as a measure of personal safety, TRCF was inadequate. What really mattered was that companies avoid killing people. It turns out that the hazards that give rise to most injuries (for example, slip and trip hazards) are not hazards that typically kill people. This means that focusing on the hazards that are producing injuries, in order to drive down the injury rate, does necessarily reduce the fatality risk. The obvious solution is to use *fatality* rates in bonus systems. Where companies are experiencing a significant number of fatalities it is highly desirable to treat fatality rate as a critical safety indicator and to specify reduction targets in company scorecards and in the performance agreements of the CEO and executive leadership. However for many companies the number of fatalities is too small to be able to speak of a rate. This was the case for all the companies in our sample.

To deal with this issue, some companies chose to treat the absence of any fatality as a gate, a pre-condition that had to be met in order that the bonus be paid. Putting this another way, in the event of a fatality, the bonus for the year would be zero.[7] This was said to be a way of conveying the seriousness with which these companies were taking safety. This was evidence that they did not put profit before safety.

However it turned out that in no case[8] was the whole bonus at risk, only the safety component. Hence, even though the occurrence of a fatality did have a significant effect on the total bonus, the component of bonus related to commercial performance remained unaffected. In the end, therefore, the "fatality gate" strategy did not ensure an appropriate balance between profit and safety.

In any case, there is considerable controversy over whether it is desirable to treat fatalities as gates in this way. The argument in favour is that it conveys a sense of the seriousness of such an event. The argument against is that fatalities may be so rare that they are best seen as random events and that it is not appropriate to penalise people in this somewhat arbitrary way that ignores whatever other good work may be going on in relation to safety. Moreover, if a fatality occurs early in the year, and people know that it has blown the safety component of the bonus for the year, they will be demoralised and perhaps less willing to continue striving to improve safety.

Interestingly, several companies had *financial* gates that could block the payment of annual bonuses. In one case, for example, profit had to exceed that of the previous year. If it didn't, no bonus was paid. Importantly, it was not just the financial component of the bonus that was at stake; it was the whole bonus. No matter how good the safety performance, no bonus was paid if the financial performance did not warrant it. The asymmetry is obvious. Failure to pass through the financial gate could wipe out the entire bonus, while failure to satisfy the zero fatalities requirement meant only that the relatively small safety component of the group score was set to zero, with a relatively small impact on the overall bonus. It is hard not to read this asymmetry as a statement about the relative importance of profit and safety in bonus systems.

Process safety indicators used in the group component

When it came to process safety, our sample of companies had made very little progress in developing metrics for inclusion in the group component of bonuses. A few companies were counting Tier 1 process safety events, as defined in Chapter 3, and one had developed a second metric concerned with the timeliness of safety-critical maintenance. Apart from this, process safety was being dealt with in a highly subjective manner. For example, one company specified merely that the target was to achieve "improvements in process safety". Some companies had made impressive progress in developing sets of process safety metrics, but such process safety scorecards are not easily incorporated into group multipliers (see Chapter 7). At best, remuneration committees considered these scorecards as qualitative input into their deliberations about an appropriate group multiplier.

One company acknowledged this subjectivity in an interesting way. It did not have a group safety metric as such, but rather, what was called a "sustainable development" indicator, represented as "TRCF+". TRCF was, of course, the total recordable case frequency, and the plus represented an adjustment made by the CEO to take account of "health, security, safety (process and personal), environmental and social performance". So, in one recent year, the TRCF was 1.26 which was adjusted down to 1.13 in light of these other components of sustainability. The symbolism of TRCF+ is highly significant. It encapsulates the pre-eminence of TRCF as the primary indicator of, not just safety, but more generally HSE (health, safety and environment) and even social performance. It also symbolises the fact that all other elements of HSE, in particular process safety, are incorporated into the indicator in an essentially subjective and ad hoc manner. We were told that process safety would only affect the final result in a perceptible way if the company suffered a catastrophic incident.

Another company explicitly itemised the factors that its remuneration committee took into account in evaluating HSEC (health, safety, environment and community) performance in one recent year. They were:

Three fatalities;
Improvement in TRCF;
Positive outcomes in respect of our endeavours with community;
Solid performance in HSE risk management, occupational health and environment.

On the basis of these factors the remuneration committee decided that the HSEC result for the Group was "marginally above expectations".

The inherent subjectivity of this evaluation is obvious, as is the emphasis on personal as opposed to process safety. Process safety is not mentioned at all, although it is potentially included in "HSE risk management".

To summarise, process safety played very little part in determining the group score for the purposes of bonus calculations. Moreover what little impact it may have had was quite invisible because of the highly subjective ways in which the final scores were determined. For most companies, then, the group component of the bonus played little or no role in directing attention to process safety.

The relative importance of safety in performance agreements

So far we have considered the relative importance of safety in group scorecards. We look now at the place of safety in individual performance agreements.

Although very few individuals can or should be held accountable for Group level metrics, there are many senior line managers, responsible for smaller business units and assets, who can indeed be held accountable in their performance agreements for the safety performance of the groups they lead. This was in fact being done in our sample of companies. For example, a plant manager's performance agreement we saw included injury rate targets as well as targets for various process safety indicators relevant to his plant. We discuss this in more detail in Chapter 7. On the whole, however, injury rate and loss of containment targets cannot usefully be included in the performance agreements of ordinary employees.

The situation is different for lead indicators, that is, indicators of safety-related activity.[9] Where employees are expected to engage in routine and repeated safety-related activities, these can be used as safety metrics. For example, if individuals are expected to do safety observations on each other, or have regular safety conversations, then target numbers can be specified and included in performance agreements. Or if senior managers are expected to participate in a certain number of incident investigations or safety meetings, this too can be specified in performance agreements. We observed this pattern in some of the companies we studied. However a feature of such indicators is that the required *quantity* can be achieved by sacrificing *quality*. It is easy enough to build up one's numbers of safety conversations by having brief and superficial conversations. This is not necessarily dishonest; it is simply the most practical way for busy people to ensure that they meet their targets, but it undermines entirely the value of the activity.

It is far better if individual performance agreements contain safety-related goals or projects, as we discussed in Chapter 3. For example, one senior manager's performance agreement that we saw included the following goals: "putting safety critical maintenance back on track", and "improving reporting and analysis of incidents". For more junior employees, one goal we saw was "assist with resolving HSE issues". Now the interesting thing about all these goals is that the question to be asked in a performance conversation is not, "have you complied with this?", but rather, "how well have you complied with this?" or "what did you do to comply with this?", questions that require more reflective and evaluative answers. This reduces the risk that compliance with safety-related goals in performance agreements will be merely token compliance, as can so easily occur with goals that are expressed quantitatively.

In some of the performance agreements we saw there were some innovative ideas worthy of note. We mention, in particular, activities that were specified in the performance agreements of a finance manager and a commercial manager. These were particularly interesting precisely because such people are often thought to have little to contribute personally to safety, whereas in fact they are very influential. The commercial manager's performance agreement specified that he would set up a committee to raise awareness of HSE in his group. Out of this committee came the idea to hold a half day process safety workshop for his staff in which they examined a number of major accidents around the world. The aim was to identify things that commercial people could do to reduce safety risk. As he said during the interview,

> I think this was really powerful because it made people realise how important commercial is, because we actually set up the ventures and operating agreements and if you like we set up what the expectations are for the way we will work from day one and it will impact many hundreds of people for potentially 30 to 40 years based on what we put into those original agreements.

The performance agreement of the finance manager specified three things of note. The first was that, in all dealings with others, he and his team should specify that safety comes first. Although this might appear platitudinous, it was surprisingly important, he said

> One of the things I've learnt is that if a finance person says that safety comes first, people believe it, because their view is that finance controls the purse strings and if finance says that safety comes first, then it must be true.

The other two elements of note were to ensure that when his staff were dealing with customers and with contractors, they discussed not only cost but also safety. They needed to convey the message that cheapest was not always best and they needed to demonstrate that they were willing to pay for safety.

Finally, we have seen performance agreements for top business leaders that specify that they should read certain books on process safety, require their direct reports to read these books, and discuss with them what they had read.

These are all excellent initiatives. They demonstrate the real potential of performance agreements of top managers to make a difference with respect to safety. Moreover a number of these objectives dealt implicitly or explicitly with process safety as well as personal safety.

In fact, however, in the performance agreements we looked at there was relatively little reference to non-quantitative safety-related goals and projects, such as those just described. The main focus was on contributions to the organisation's business objectives.

Finally we should comment on the weightings attached to safety in individual performance agreements. Some performance agreements made no attempt to specify weightings. One company left it to employees individually to specify weightings within certain allowable ranges, as follows:

Business	25–75%
HSE	15–25%
Organisation and employee development	10–25%
Behaviours	10–25%

These ranges suggest that safety is to be a relatively small component in the overall performance agreement.

In fact, however, many employees failed to specify any preference in this regard. When questioned about this they said they would simply accept the company's default option, although they were unaware of what that was.

None of this is surprising. As we have already noted, the performance evaluation process is inherently subjective and seeks to come up with an overall assessment. It would be virtually impossible for a supervisor to factor in to such an overall evaluation a precise numerical weighting for safety (or any other component).

All of this raises the question of what really are the criteria that determine the ultimate evaluation – what is it that supervisors regard as really important? We return to this question in Chapter 6.

A remarkable initiative

Let us return now to the question of fatality risk. Treating fatalities as gates, as described earlier, is a rather ad hoc way of giving greater weight to fatality risk. One company in our sample – we shall call it company A – had taken this issue very seriously and had developed a far more sophisticated strategy for emphasising fatality risk. It is worth laying this out in some detail.

Company A realised some years ago that it needed to focus on fatality risk as opposed to injury risk. To do so, it made use of a semi-quantitative risk assessment process based on the informed opinions of groups of knowledgeable employees. The starting point of the process is to identify all potentially fatal hazards at a site and to recognise that there are likely to be many more near misses (initiating events) associated with these hazards than actual fatalities. Based on its own experience, the group estimates the frequency of these near misses. For example, near collisions on mine sites between large haul trucks and light vehicles might occur once a week. The group then estimates the frequency with which these near misses result in contact (say 10%), the frequency with which these contacts result in injury (say 50%) and the frequency with which these injuries result in death (say 1%). Multiplying these together gives a fatality frequency of about .025 per annum or about 2.5 deaths per 100 years. Next, the group identifies possible controls, the reliability of these controls is rated, based on informed group opinion, and for each control (or combination of controls) the fatality risk is re-estimated. Some controls give greater risk reduction than others and the group is thus in a position to recommend appropriate risk reduction initiatives. All this is admittedly very rough and ready, and likely to be out by orders of magnitude. Moreover, it is not readily applicable to process hazards, which generally require much greater technical analysis. Nevertheless the methodology gives people a better understanding of some of their most significant fatality risks as well as insight into relevant risk controls.

Company A's experience with this approach has extended over several years. In the first years it simply required that all sites go through this semi-quantitative risk assessment process, identify risk reduction measures, and implement them. The extent to which sites had reduced their fatality risk in this way was assessed and included on the company safety scorecard, thus affecting bonuses. In this way senior executives for the first time had an incentive to focus on fatality risk.

Subsequently, the focus shifted to ensuring that the controls identified in this process were actually in place. They were described as critical controls and sites had to develop critical control monitoring plans (CCMPs). These plans had to be validated, that is, shown to be working, and the degree of compliance with this requirement became an element in the safety scorecard.

The following year saw an improvement on this validation process. It was recognised that the data on compliance were too superficial and that what was required was that site leaders should be actively engaged in monitoring the effectiveness of critical controls. For business units to score "very good" on this criterion, business unit leaders would need to be able to personally vouch for the fact that site leadership engagement had been "very good". This personal assurance was to be part of the business unit leader's personal performance agreement.

Table 5.3 gives a flavour of the CCMP in practice. This was a sign outside an acid loading facility at a company site that we visited. On the left side are the site's most significant fatality risks and on the right side, the critical controls. The CCMP for this site would involve monitoring these controls.

One of the virtues of this sign is that it identifies the critical controls for anyone who is seeking to monitor them, making it relatively easy for leaders to engage in this task. One of the limitations of the process revealed by this sign relates to the hierarchy of controls. It is widely accepted that PPE (personal protective equipment) is one of the last and least reliable lines of defence that is only needed if other defences or barriers have failed or are absent. For instance strong acid containment PPE is only necessary if acid has escaped from its containment tanks or piping. Thus, one of the fundamental controls for protecting people against contact with acid must be to ensure the integrity of the containment systems. This is missing from the above list, yet we would surely want leaders who were monitoring

Table 5.3   Critical controls at one Company A site

TOP CRITICAL RISKS	
Hazard	Control
1. Significant exposure to strong sulphuric acid	Restricted access; Strong acid containment PPE
2. Mobile equipment interaction	Visual contact (8 metre rule) High vis vest; walkways; rail stops
3. Fall from height	Handrails, guardings; Fall protection when required (training)
4. Struck by object	Restricted access on loading rails; Barricades overhead work

critical controls to be asking questions about the integrity of the containment systems.

Alongside this focus on Critical Control Monitoring Plans, Company A had a second strategy for introducing fatality risk into its bonus systems. It had developed a system for reporting Significant Potential Incidents (SPIs), and introduced incentives into its bonus arrangements to increase reporting. Over a three-year period the number of reported SPIs increased from approximately 300 to approximately 1400, testimony to the significance of the bonus in this case. Once the system of reporting had matured, the next step was to ensure that lessons from these incidents had been properly learnt. The company therefore introduced targets for numbers of action items stemming from the analysis of these incidents, as well as targets for the close-out or completion of these items. It then realised that too many of the action items were being closed out using controls too low in the hierarchy, that is, without consideration of more fundamental engineering controls. It therefore instituted a regular audit of the quality of the corrective actions taken in response to SPIs.

A third way of focusing on fatality prevention was to emphasise major safety audit findings. There was a threshold requirement that there were no repeat findings, that is, that the company acted on the audit findings the first time round. Beyond that, the score was determined by time to closure of audit recommendations.

The relative weighting of these various components is presented in Table 5.4. The first thing to note about this table is that the scores at different points in the organisational hierarchy are different. In particular, despite all that has been said above, the only safety metrics relevant to the CEO and his direct reports are two injury rate measures, the all-injury rate and the rate of slightly more serious injuries – lost time injuries. There is no reference to fatality risk controls at all in their scorecard and hence in their bonuses! How can this be? It turns out that shareholders were sceptical of the metrics related to fatality risk, regarding them as too subjective, and demanded that the CEO and his direct reports be evaluated using more objective measures. Unfortunately, the only possible measures were injury rates. Hence Company A is right back where it started with respect to its most senior executives, rewarding them for injury rate reduction, but not for reducing fatality risk. As far as we are concerned, this is a serious flaw in the whole scheme. It means that top executives will continue to be most interested in injury rates. And given that "what interests my boss fascinates me", this

Table 5.4   Safety scorecard for annual bonus in Company A

Safety Components	Weightings		
	CEO and exec	Senior mgrs	Lower mgrs
Lost-Time Injury Frequency Rate (LTIFR)	50%		
All-Injury Frequency Rate	50%	40%	20%
Critical Control Monitoring Plans (CCMP)	0	25%	35%
Significant Potential Incident (SPI) Closure Rate	0	20%	25%
Major safety audit findings – no repeats and time to closure	0	15%	20%
Total	100%	100%	100%

emphasis will percolate downwards, tending to undermine the intent of the remainder of the scheme.

There is one qualification to be made to this pessimistic conclusion. For the senior executive group the safety score is subject to a "fatality adjustment". This is not a gate, as described earlier, which would lead to a zero safety score in the event of a fatality or fatalities, but rather a reduction in the score, according to some predetermined formula. In this way it is hoped that the top executive group will maintain an interest in the fatality risk reduction measures even though they are not directly incentivised to do so.

If we look now at the remainder of the table, the emphasis is clearly on measures aimed at reducing fatality risk. For both senior and lower level managers the bulk of their score is made up of such measures, while the injury rate measure contributes less than half the final score. For lower level mangers the emphasis on fatality risk reduction is even greater. The reasoning was that, as one moves down the organisational hierarchy, managers have decreasing control of injury rate metrics. Incentives therefore need to be focused on matters that are to a greater extent under their control.

Leaving aside the top executives, therefore, the scheme laid out in Table 5.4 is a major step forward in incorporating fatality risk systematically into the bonus arrangements. It is for this reason that we have devoted so much attention to it here.[10]

To complete the picture we need to know how the safety scorecard is incorporated into the total group scorecard. The system is additive and safety counts for 25% of the total group scorecard, the remaining 75% being made up of commercial indicators. So, despite the admirable focus on fatality risk, company A does not stand out from the other companies in our sample in terms of the overall emphasis on safety in its annual bonus structure.

We note finally that this discussion has not featured a distinction between fatality risk and major accident risk, because to date Company A has not had a focus on major accident prevention. It is now beginning to rectify this lack, and has constructed the diagram in Figure 5.1 to represent the situation.

This triangle is conceptually useful. It not only distinguishes between fatality risk and injury risk, as discussed above, but it also

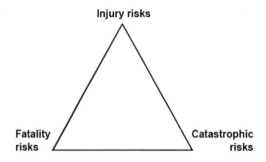

Figure 5.1   Three types of risk

distinguishes between fatality risk and the risk of catastrophic events, or major accidents. This is a distinction that is missed in current discussions about personal and process safety. There is a tendency in such discussions to treat personal safety as secondary, perhaps even trivial. But the fact is that personal safety hazards can be fatal, the most obvious one being driving. The triangle helps us keep sight of this fact.

This new style of thinking led Company A to re-evaluate its method of semi-quantitative risk assessment. The method relies heavily on "experience around the table". However process risk may not have resulted in a catastrophic outcome in the experience of those around the table. There is therefore a need for a different strategy altogether for identifying critical controls for process safety risks. In part, this means that risk analysts must look at accidents that may have happened in similar plants in other companies or elsewhere in the world. It also requires a more technically sophisticated understanding of the hazard in order to be able to identify critical controls, or barriers or defences, to use the language of Chapter 3. In theory, company A's fatality risk-reduction strategies could be modified to deal with the risk of major process accidents, as well as more conventional fatality risks. In practice, this has not yet happened. The challenge, as Company A recognises, is to extend its framework of CCMPs and SPIs to incorporate process safety hazards. API 754, discussed in Chapter 3, is relevant here. Tier 3 and Tier 4 indicators are concerned with the performance of critical controls for process safety risk. We develop this further in Chapter 7. Company A will need to follow this path.

Finally, it should be acknowledged that many companies in our sample were focusing on fatality risk in various ways. One way was to identify the most significant fatality risks – for example, driving, entry into confined spaces, walking under suspended loads – and to develop a set of simple rules to deal with them called, for instance, "lifesaving rules", "golden rules", or "cardinal rules". Any violation of these rules is met with immediate consequences. However none of these companies had systematically incorporated fatality risk into its bonus system as Company A had.

Conclusion

We conclude this chapter by summarising the potential effect of short-term bonus systems on safety. Short-term bonuses generally had two components, relating to group and individual performance. However, "ordinary" employees, that is, most employees, have no capacity to influence group outcomes. Prima facie, therefore, there can be no justification for including group measures in incentive schemes for such employees. No company we studied was able to provide any justification for this widespread practice, other than that it served to "align" the interests of the employee with those of the company. It is true that linking individual pay to the fortunes of the company aligns the interests of employees with those of the company, but it is hard to see what practical consequences this can have, precisely because ordinary employees can have no perceptible influence over company fortunes. In fact, paying bonuses to employees on the basis of group performance looks surprisingly like profit-sharing for its own sake, a totally different proposition. If that is what the designers of bonus systems intend, perhaps they should be explicit about this.[11] Of course, the most senior managers in any company *can* influence group performance measures by their actions, and it make sense to incorporate group measures in the bonus schemes for such people.

Moreover, for the group component, safety counted for 20% on average, with the remaining 80% determined by commercial indicators. This meant that safety was at risk of being swamped by commercial considerations. This problem is exacerbated by the fact that remuneration committees treat all numerical scores as guidance only in arriving at a judgement of overall performance. Under these circumstances it is not possible to ensure that safety is given the weighting it nominally receives. Only one company had developed a strategy to avoid this problem. Finally, we found that process safety disappeared from view in most group scorecards.

As for individual performance, performance expectations are set out in individual agreements. For the most part these expectations amount to tasks or projects, not metrics, and individuals can plausibly be evaluated on how well they have achieved those goals. Bonuses tied to these individual performance objectives therefore have at least a potential to motivate the required behaviour. In practice, however the requirement that people be ranked on a bell curve creates considerable disillusionment which tends to undermine the potential incentive effects of the performance review process. This will be taken up in Chapter 6.

Safety is generally only a small element within performance agreements. On the whole, therefore, the potential for performance agreements to highlight safety-related objectives is not being realised. On the other hand, group safety metrics, including process safety metrics, were indeed included in the performance agreements of many high-level line managers, and in addition we observed some innovative ways in which safety, including process safety, had been effectively incorporated into the performance agreements of some other senior managers without line responsibilities.

Notes

1   This description is true for companies that used an additive system (see below) as well as for those using a multiplicative.
2   BP Annual Report and Form 20-F, 2012, p. 132.
3   Ibid.
4   Another four-point scale we encountered was: outstanding, commendable, effective, needs improvement. An example of a five-point scale is as follows: exceptional, outperforming, fully effective, requires improvement, requires fundamental improvement.
5   Hopkins, 2008; Hopkins, 2012.

6 A recordable injury or illness is one that results in death, days away from work, restricted work or transfer to another job, medical treatment beyond first aid, loss of consciousness, or any other significant injury or illness diagnosed by a physician or other licensed health care professional. This is a definition used by the US Occupational Safety and Health Administration 29CFR 1904. OSHA requires all such cases to be recorded and reported.

7 This might not apply if the fatality was seen to be totally beyond management control, such as an off-site traffic accident resulting in death. According to one company, "The extent of the financial impact of any fatalities on the STIP score for executives is based on a judgment process that assesses the impact of leadership, individual behaviour and systems in the incident."

8 With possibly one exception.

9 Hopkins (2009) has discussed the distinction between lead and lags indicators and found it to be ultimately meaningless. However we use the terms uncritically here because they are so widely used.

10 Since writing this we have discovered that company A has retreated further from the use of fatality risk controls in bonuses for senior managers. We do not know the reason.

11 Perhaps the argument is that profit sharing promotes good will. However there is a risk that such payments will come to be seen as entitlements. We were told that employees in one company had sought legal advice when their bonus was cut one year due to poor Group performance. The poor performance was in one part of the world and employees in other parts of the world did not see why they should suffer the consequences. For them, the failure to pay a bonus became a source of ill will towards the company.

Chapter 6
The Impact of Incentive Arrangements

Chapters 4 and 5 looked at the structure of incentive arrangements in a somewhat abstract way. In this chapter we focus on how people respond to incentive schemes in practice. The chapter is based on a series of in-depth interviews we conducted with senior managers in a subset of our sample companies.

One of the striking findings of this research was the ambivalence of informants about incentive schemes. This was particularly surprising, given the huge level of resourcing devoted to them. Companies spend months every year deciding on appropriate payments for each eligible person; metric design and performance measurement consume substantial corporate energy; and the bonuses themselves can dwarf employees' fixed remuneration, particularly for top managers. Nevertheless, while some senior managers we spoke with were confident about the motivational capacity of the incentive arrangements in their companies, a much larger group was not. These latter interviewees rejected the notion that they could be making decisions guided ultimately by their performance agreement and its promised financial rewards, particularly when it came to safety. It seems to us extraordinary that corporations might be wasting their energy on complicated and expensive incentive arrangements that simply do not serve them.

Accordingly, in this chapter we investigate the impact of these bonus arrangements through senior managers' accounts of what guides their daily decisions, and their experience of the incentive and performance management processes. We attend to three elements of incentive arrangements – the performance agreement, the performance evaluation and the financial bonus itself – to examine what drives decision making. We look also at unintended consequences. Finally, we examine the claim that safety is special, being a fundamental value that managers hold, and that financial bonuses are therefore unnecessary as a means of encouraging safety-conscious decision making. It is important to consider this claim because some of our informants used it as a justification for imperfect safety metrics and for the low weighting given to safety in their company's incentive arrangements.

We draw the reader's attention to one particular limitation of our work – there are no CEOs among our interviewees. This came about because we were in the hands of senior safety managers when it came to the selection of interviewees, and they were not able to obtain interviews at the CEO level. The result is that any comments about CEO motivations are based on the perceptions of our non-CEO respondents. We did however interview some very senior people, including heads of national operations within global corporations.

One other reservation is that the number of companies involved in this second stage of interviews was small. We cannot therefore be certain that our conclusions are valid for all the companies in the larger set. However there is no a priori reason to think that they are not.

The content of the performance agreement

In Chapter 2, we cited evidence that suggested that incentives were motivational, in part, as a result of their connection to evaluations against performance agreements. We provided an example from one of the inquiries into the blowout in the Gulf of Mexico, which found that individual performance agreements often specified that employees should contribute to cost reduction goals. The inquiry found that of 13 employees whose evaluations it examined, 12 had documented ways in which they had saved the company large sums of money. From this we concluded that we could expect the content of performance agreements to have an effect on the decisions of employees. We address this hypothesis here empirically.

One of our most interesting findings was that, for most managers, their individual performance agreement was a restatement of their job description and the business and safety objectives of the team or group and, as such, gave no additional guidance as to priorities. Our informants claimed that they do things because it is their job and that, for the most part, their performance agreement reflects their job description. This was captured strikingly in one interview. We asked this senior manager about his priorities and role, which he described in detail. When then asked specifically what was in his performance agreement, he responded: "It's basically what I just told you." Another explained it like this:

> I believe that a lot of this stuff is put in place to make sure that managers do what they're supposed to do: have conversations with their employees, meet the targets of the organisation. I believe I do those things, [so if I had checked off the items on my performance agreement] I would have gone, 'yeah, I've met that, I've met that, I've met that'. But these are the key things that I'm going to do anyway. I don't go back to it on a daily basis. It's not a task list.

Because the performance agreement was viewed as a restatement of what managers already understood to be their job, most stated that they did not look at their performance agreements regularly, evidence that it is not the document itself that is driving their priorities and decisions. Here is another comment to this effect:

> We produce these KPIs ... I then put that in the draw and don't worry about it. I get on with the job. They reflect the job. If I'm not doing my job properly then that is going to come out on its own.

The performance agreement, according to this account, is not a relevant document in the context of daily decisions. From the perspective of this manager, this is how it should be.

While these claims are common, we found that the performance agreement and the requirements of the job were not always the same. One of the reasons for this inconsistency was that, in a dynamic work environment, priorities and jobs change, but performance agreements tend not to be updated to reflect this. For example, one senior manager's job had significantly changed since the development of his performance agreement, as had the priorities of the organisation and expectations about project delivery. As such, his evaluation was against a largely separate and tacit set of criteria. An almost complete change such as this was not common, but some change in priorities was generally considered to be normal. New standards and policies might need to be implemented. An example of this was a new driving behaviours program that one company had decided to implement, before the next round of performance agreements was due to be drawn up. The manager was expected to implement this although it was invisible in his current performance agreement.

We discovered, quite surprisingly, that at times there were significant differences between performance agreements and real job priorities that the managers themselves seemed unaware of. An example of this that appeared time and again in interviews was the deviation between 'safety' as understood by managers, and 'safety' as it was measured in performance agreements. One senior manager described in great detail his efforts in major accident prevention, including the development of safety cases and incident simulation training. However, neither of these tasks appeared in his performance agreement. Rather, safety, according to his performance agreement, was measured in terms of injury rates and other personal safety metrics. Similarly, in another senior manager's performance agreement, personal safety was measured in terms of injury rates, however, the manager actually prioritised prevention of permanent injury to his workers. Another spent a lot of his time on welding quality assurance, which has significant implications for asset integrity, yet it was not listed in the performance agreement. These managers and others like them were often not explicit about the gap between where they focused their attentions and what they had formally committed to in their performance agreement. But when it was pointed out, they generally accepted it as the way things were. This raises the question: why were these managers not more concerned about such a gap when, theoretically, their performance evaluation and bonus rides on achievements against their performance agreement?

The answer is that, remarkably, individual performance evaluations often do not turn on the details of the performance agreement. A clear illustration of this is in the perceived impact of injury rate outcomes on performance evaluations. Nearly all of the performance agreements that we have seen include injury rate measures not only in the group, but also the individual component. Most people did not perceive this metric to be within their control, though it was common for managers to state that they were ultimately 'accountable'. Despite this theoretical accountability some of our informants explained that failure to achieve these targets had no discernible impact on the final performance evaluation and bonus. Rather, they believed that their final evaluation was more qualitative, and perhaps even unpredictable. As one explained:

> I don't see that individually they're going to hold me too accountable for [the injury rate], but there are other things here that they might, I don't know, it does turn a little bit on the individual ... At every performance review I've ever had [the agreement] has always got pulled out but I don't believe that anybody has ever scored me.

All this suggests that the performance agreement itself plays a relatively small part in the performance evaluation process. That process will be discussed further in the following section.

In addition to perceptions that the performance agreement overlapped with 'the job', as well as a general sense that employees may not be held strictly accountable for its targets, another issue that we found was that performance measures were too numerous to effectively influence behaviour. Of the samples we have seen, individual performance agreements include 20 or 30 items under four or five subheadings, in addition to a series of more general statements about a manager's areas of responsibility. To illustrate this complexity, we have developed a fictitious performance agreement based on the elements we observed in actual performance agreements – see Appendix 2. Readers are referred to this appendix to get a "feel" for what we are talking about. Some managers believed this large number of items was manageable and reflected the nature of their role. However, others were more critical and regarded such lists as unmanageable. They listed two or three, sometimes up to five key priorities that they could attend to throughout the year. Similarly, when asked what their manager cared about, the list was less than five.

It is also worth noting that the four or five categories in individual performance agreements could be weighted differently. In one company it was left to employees to assign a weighting to different components (such as business KPIs, safety, employee mentoring). However, employees in this company commonly did not bother with this step in the development of their performance agreement. Again, this tells us that the precise performance measures and their weightings were often considered by our informants to be unimportant.

The preceding comments reflect the experience of the majority of managers we interviewed. There was however a minority of managers for whom the situation was substantially different and who, in particular, claimed that their performance agreement had an impact on their daily decisions. The managers for whom the performance agreement was a useful document, on the whole, had indicators that could be described as SMART: specific, measurable, assignable, realistic and time-related.[1]

To put this in context, many of the measures we saw in individual agreements failed the SMART test. To use the injury rate example again, on several occasions we observed that individual performance agreements included an injury frequency rate target of zero. This is specific and measurable, but it is not a realistic goal that can be achieved by a manager across a business in any 12-month period. Where it was included, it was aspirational, and in fact discounted by our informants.

To take another common example, individual performance agreements tended to be dominated by financial and business metrics. These primarily included high-level measures of financial and business outcomes, such as expenditure, net production and project milestone delivery. These metrics were not seen as wholly unrelated to the job, but it was commonly acknowledged that they did not offer the necessary granularity to be useful on a daily, weekly, or even monthly basis. Moreover, they commonly measured outcomes that might be outside any one manager's control. Interestingly, however, people in this position typically accepted that they remained 'accountable'. Metrics such as these are measurable, which is part of their appeal. They are also theoretically assignable, in that particular managers can be held to account for them. But they are not realistic if they do not in fact lie within the control of the manager concerned.

These two examples – injury rates and business metrics – are part of a broader trend in which performance agreements tended to be filled with numerous role statements and targets that left them opaque and chaotic. This undermined their usefulness and even credibility.

The managers who took their performance agreements most seriously had indicators that were SMART. They referred to these indicators on at least a weekly basis and had operationalised them into their daily work. As one site manager explained:

> We review the indicators weekly because I'm not going to look at this once a year and say, 'Wow, that was all off track for the whole year'. What we can measure, we measure live. So, for example, we can measure our alarm rates. I know now exactly the alarm rate outside is one alarm per hour per console. I can see that live on my desk. And I can also click on it and see across the plant which consoles are contributing in what weighting. So immediately if that thing turns red, I can head down there and say, 'What help do you need? What's going on here? What is sitting underneath that number?' Fast information leads to fast intervention. As a leadership team, we have a scorecard that builds up over the week and every week we'll review all those metrics formally so that they're not left to drift.

For this senior manager, the metrics in his performance agreement were specific, timely, measureable and assignable. Because of this, it made sense for him to use them as foundations for his daily and weekly management. Importantly, the precise metrics he would ultimately be evaluated against were not meaningless, but indicators of potential issues that he, as the site manager, could influence personally and through his leadership team to achieve the company objectives.

A common practice among managers who actively used their performance agreement was its translation into another, more visible format. One manager had summarised it on a sheet of paper broken into quadrants. This document was often if not always carried by the manager, and was pinned up in his office. Another translated his performance agreement into a series of scorecards, which he, too, carried with him. For each of them, this translation simplified their agreement so it was useable. It was always with them. It was viewed as important.

In summary, most of our informants felt their performance agreements were of little importance. This was because agreements were understood to be a restatement of the job description, because they included too many indicators that failed the SMART test and, as we will discuss further in the coming section, because the performance evaluation often did not turn on the details of the performance agreement. However, this is not a reason to abandon performance agreements. A minority of managers were explicitly using their agreements to guide their actions. Performance agreements, in other words, can be worthwhile, if they are designed with care and in particular if any indicators they use are SMART.

Impact of the performance evaluation

A key aspect of the incentive arrangement is the performance evaluation, which most often involves a one-on-one discussion between a supervisor and an employee about their performance over the previous 6 or 12 months. For many people that we spoke with, this was the most significant aspect of the incentive and performance management process because it gave an indication of how much they were valued as employees. Such judgements would also have a bearing on their future careers. For some managers it was the conversation itself that was important but, more commonly, they were seeking an above-average performance appraisal. One senior manager explained the impact and significance of this process as follows:

> The financial bonus doesn't make a lot of difference to me. Getting 'outstanding' for me is going to help where I'm going to go next [my next promotion or job]. If I was graded 'satisfactory' or 'needs to improve' then I'm not very happy. This evaluation is important to me in terms of supporting career progression and in terms of pay rises.

This account of the process was very common. Our informants wanted to be judged above average, not specifically because of its implications for the annual bonus but because it tapped broader motivations. This finding is in keeping with our earlier argument that we could expect incentive schemes to be motivational for a host of reasons, among them the need for approval, the need to belong and the need to be recognised as making a valuable contribution.

In the previous section, we explained that the performance evaluation does not turn on the performance agreement itself. Rather, it is a qualitative judgement of overall performance. Managers described getting "hit hard" based on performance in a particular area which sometimes was not even in the performance agreement. Emphasising this, a manager reflected: "I know people at quite senior levels who don't even look at [the performance agreement] when they have discussions with their people." This makes the evaluation process all the more significant. It is the commentary by the supervisors during the performance evaluation that provides the employee with the clearest indication of what is really important. This is perhaps a little ironic. The whole scheme of performance agreements is intended to convey priorities and yet it is really the response of the supervisor that most effectively determines those priorities.

Here is one example relating to a site manager. Safety was one of the objectives specified in his performance agreement, but it was only one of many, and there was no real indication in the performance agreement of the priority to be attached to it. A conversation with his supervisor changed everything. Here is how that site manager reported the conversation to us:

> The discussion was a real learning point for me – where the boss said, 'You know Bill, the safety performance is not where it needs to be. That's all about your leadership. You haven't done enough'. And I can remember arguing and saying, 'Well tell me what I should have done differently'. And he said, 'Bill, that's not what this is about. This is about outcomes. The outcomes were not where they needed to be. Don't ask me what you should have done. You were there'. This was an important leadership milestone. I realised, 'Yeah, OK, this is about me'. And I asked myself 'Could I have done more?' Yeah I could have. 'Could I have spent more time in the field? Could we have organised more assurance? Could we have done better interventions?' Yes we could have. It wasn't easy at the time but it was probably a defining moment. I didn't like hearing it at the time but he (the supervisor) landed it well.

After this conversation, he said, safety was always a high priority among all the tasks specified in his performance agreements. One can easily imagine other conversations that would indelibly etch other priorities in the minds of those involved. In this case the priority happened to be safety. Moreover, one imagines that for the supervisor in this case *his* priorities came from a similar performance conversation with his boss, rather than from his performance agreement as such. Insofar as this is so, it would seem that the most important component of the whole performance management process is the performance evaluation itself, because it is really only in these conversations that priorities are clearly articulated.

A second aspect of the performance evaluation that has far-reaching consequences is that the evaluation is relative to others. The problem is that most people must be graded as average. This means that relatively few people can get the above average appraisal that they would like and which they would see as positive feedback.

Managers were aware, too, that across the organisation most managers were high performers relative to whole populations, which made an average grading all the more difficult to take. Some informants claimed to take the whole process with a grain of salt, but there was evidence that the formal categorisation of individual performance was inhibiting the efficacy of incentives. Our conclusion is that the relativity of the evaluation process at best undermined the potential motivational impact that could come from the process. At worst, it was demotivating. Here is what one of our interviewees had to say on these points:

> If you're a person who has a fair bit of pride about their performance, you hold yourself to account, and you expect to get [evaluations] that are north of [average] and you're really not satisfied unless they are well north of [average], and then you get [below average], which is basically on the verge of being put on performance management, then it's quite jaw dropping. For me it's not about that number multiplied by another number equals some financial outcome. It's more the fact that this is what [the company] thinks of me. And I think by and large that's what most people's view is … This is the one time of the year that line managers are forced by the process … to differentiate the staff and to put them in a line from the highest performer effectively through to the lowest performer.

A senior manager with a career's worth of experience in administering remuneration schemes captured this impact:

> People say that they are not money motivated but blimey if they don't have it … particularly in a relative sense. Suppose you've got two guys: one goes above and beyond the other, while the other slacks. If the hard worker gets a $20,000 bonus, they're really happy, but if they come into work tomorrow and find out that the other guy got $19,000 then you've got a problem.

The sense here is that the direct financial incentive matters, but in addition to that, the motivational impact of the bonus turns on the feedback it provides to employees on how much their performance is valued in a relative sense. If there is a perception of unfairness, as the manager quoted explained, "we have a problem". One is reminded here of the ultimatum game described in Chapter 2: person A regarded the financial benefit on offer as worthless if person B received what was perceived to be an unfairly large benefit.

There was a further problem with the whole ranking process: immediate and direct feedback was not possible because the evaluations had first to undergo the moderation process. In at least one organisation we visited, this evaluation was never given. Rather, staff ultimately figured out what the outcome must have been, based on their bonus. Many people we spoke with were frustrated with this process. It prevented them from getting and giving timely feedback, and adjustments to the initial evaluations undermined their perceived legitimacy.

In summary, while the managers we spoke with generally accepted their performance agreement and evaluation processes for better or for worse, there are clear weaknesses in the system that are worth highlighting. Given that performance measures are so numerous, they cannot clearly indicate key priorities, and guidance as to where an employee should focus attention is left to supervisors in the performance evaluation itself. In effect, this means that the evaluation is the most important component of the performance management process because it is really only in these conversations that priorities are clearly articulated. Given its importance, it is vital that the evaluation process is as transparent as possible. However, we found that the all-important feedback cannot always be given in an immediate and unmediated way. This undermined the perceived legitimacy and ultimately the impact of the performance evaluation. Moreover, with evaluations distributed on a bell curve, most people by definition must be appraised as average. This is unnecessarily deflating.

Does money motivate?

The money involved in incentive arrangements was rarely identified as motivational in the first instance. The idea that senior managers might make safety-related decisions on the basis of financial incentives seemed abhorrent to many interviewees. Instead, they said, such decisions were based on values, morals even. This question of whether safety is a special case when it comes to motivation will be addressed later in this chapter. Here, however, we press on with the motivational capacity of financial incentives. In interviews, we questioned our informants on the impacts of the financial bonus in two ways. We asked whether they might be more motivated by money if the amount was higher. What would their response be if, for instance, their bonus was 200% of base salary, which is a common at-risk component for the CEO of a multinational resource business. We also asked whether their motivation would be impacted if the bonus was eliminated (assuming their base salary was increased to remain market competitive). Our following analysis centres on these two scenarios.

200% of salary

We found that people were more likely to feel they would be motivated by the financial aspect of the bonus if a greater component of their salary was at risk. Interestingly, however, no one we spoke with felt that *their* bonus was substantial enough to influence their priorities and decisions. It was only *others* with higher bonuses who might be at risk of being money motivated. Here is one statement along these lines from a senior manager:

> I am not going to mention figures but I certainly know [my boss's] bonus. And that's a very substantial bonus which makes him want to deliver no matter what, and which influences as far as I'm concerned his decision-making process, which is not always best for the job. But my bonus isn't high enough to affect my own decision-making processes.

This senior manager is expressing the view that the decisions and priorities of *others* with a higher at-risk component of their remuneration can be influenced in this way, while his own bonus is not high enough to have this effect. Importantly, he assumes that there is conflict between the decisions that would result from the highest bonus, and the interests of the 'job'.

These concerns that financial incentives can drive decisions that are not in the long-term interests of an asset and/or company were common. As a senior manager at one company put it, they "drive the wrong attitude", sending the message that "it's more important to earn the bonus than manage the company". He gave a scenario where half the bonus rested on starting up a new asset by a particular date as a clear example of where bonuses could drive unintended consequences. He said that he would not take a position where the bonus was 200% of base salary for this reason. Another senior manager expressed a similar concern in the context of major accident risk management:

> I think the bonus element is too big … these guys would be massively incentivised to deliver on their contract irrespective of the risk they are taking. And on top of that, the risk they are taking is short term, when their decisions are actually long term [in their consequences], so there is a mismatch.

To sum up, a number of our informants were concerned that decision making by their most senior managers was being inappropriately influenced by the very large bonuses they received. We were not in a position to evaluate the accuracy of these perceptions. But the fact that such perceptions were widespread suggests that something is amiss.

Consider now the statements by our informants that financial bonuses did not unduly influence *their own* behaviour. One possibility is that they are simply unaware of the influence. Another is that they are aware of it but are uncomfortable about acknowledging what they see as the unsavoury reality. A third possibility is that the money really was not a significant consideration. There are various circumstances that support this third interpretation. The bonuses received by the managers we spoke with were typically between 30–80% of base salary, depending on how senior they were. The amounts in other words are quite large. However, due to the clustering of performance evaluations around the centre, the bonus received by any one manager did not vary much from one year to the next. Moreover, whether one ended up in the modal category (e.g. "meets expectations") or one of the categories above, did not make much difference to the financial outcome. The financial difference between categories was less than 5% of base salary, sometimes only 1 or 2%. In other words, although a significant percentage of total remuneration was theoretically at risk, this was not in practice the case. The perception of our managers was that there was very limited variation in the bonus they received. This undermined the purely financial motivational effects of bonuses.

The situation for CEOs is different. As we saw in Chapter 4, the incentive arrangements for CEOs are similar to those in the banking sector in that their bonus really is at risk. If the company does not perform as required, then no bonus is awarded at all. A win or lose scenario such as this is closer to the conditions that led mortgage brokers to focus on short-term economic gain at the expense of longer term economic stability. The experience of the finance industry in the global financial crisis was raised by some informants as a cautionary tale of where they did not want their own industry to end up. These managers believed that incentive systems were only justified if they encouraged employees to be concerned about the long-term performance of the whole organisation.

No bonus

The second question we raised was the motivational consequences of removing the financial bonus altogether. One of our interviewees was strongly in favour of this course of action. In particular he favoured removing the personal contribution to the bonus altogether, while leaving a group component that depended on company performance. His words are worth quoting at length:

> I think ranking processes are a complete waste of time personally. I think it doesn't motivate me to do anything differently and it's a monumental waste of an organisation's effort. It takes basically two months of meetings and discussions and everything with a relative tight spread of bonuses, and in discussions with individuals afterwards, they are most likely to feel demotivated … The bonus has no role in motivating me. I don't get out of bed thinking what am I going to do for my bonus next year? It's what am I going to do to make the organisation work? … My incentive is to do a good job (not to get a bonus). I don't think bonuses change behaviour.

For this manager, therefore, getting rid of the bonus system, at least that part of it that is tied to individual performance, would certainly not undermine his motivation to perform well, and would probably enhance it.

Another manager told us that career advancement was a far greater motivator than the financial bonus. Here are his words:

> I'm quite career oriented so I want to do a good job and hopefully get promotion and go up to the next level on the next project … I'm a long-term company guy, I've been more than 20 years with the company so whatever I do on this job is going to follow me through the rest of my career so I'm more about making sure this is right when it's commissioned and started up – that it sustains its start-up and runs correctly.

For this manager then the bonus was largely irrelevant as a motivator.

Finally, one of the managers we interviewed was a contractor – not a regular employee, but someone brought in to do a specific management job, who would most probably move on to another contract with another company when this one was finished. The significant point is that, as a contractor, he was not eligible for a bonus. We asked him what if anything incentivised him to deliver the job on time:

> I don't think there are any penalties [if you don't] apart from your pride, you know, and your willingness to, or your wish to do a good job and get the job over the line. But I have got no big pot of gold sitting at the end of this to drive me to get it finished earlier …. It's a small industry and if you are good at your job that will be known, if you are not good at your job you probably wouldn't last very long.

For others, however, the idea of abandoning bonuses altogether caused them to reconsider the motivational value of their own bonus. They concluded that the financial and recognition aspects of bonuses were inseparable. One senior executive captured a common sentiment well, when he explained:

> I like to be recognised for doing a great job and recognition in terms of a monetary outcome is a very attractive way of being recognised.

And even more interestingly

> I don't believe I make decisions to maximise outcomes in monetary terms [for myself]. I make decisions that I believe are right for the business and I like to be recognised for that in monetary terms.

But it is not obvious what to make of these comments. If the monetary component of recognition was removed, would he behave any differently? The implication of his comments is that he would not. After all there are other material payoffs for being recognised as having done a good job, such as promotion, which are also highly motivating. Perhaps it is simply that money is the predominant way in which good performance is recognised in the corporate environment and this is why he wants to be rewarded in monetary terms. From this point of view a financial bonus may be simply a way of communicating that recognition.

There was one other interesting argument mounted in favour of the bonus system. It was that the requirement to rank people for bonus purposes forces supervisors to have the "hard conversations" with their direct reports that they might not otherwise have. One manager told us that the absence of any bonus, would "take the edge off those end of year conversations". This claim suggests that the financial aspect is not only a tool for recognition, but also supports the provision of potentially challenging feedback.

Managing unintended consequences

One of the primary concerns about financial incentives raised in Chapter 2 was the very real potential for unintended consequences. We address this here through the experience of our interviewees.

Many of the people we spoke with rejected the possibility that incentive arrangements could have unintended consequences, either because there were checks in place, or because they had refined their schemes in response to past problems. For example, one of the known issues with incentivising safety critical maintenance is the potential for it to be ever rescheduled so nothing is listed in the system as overdue. Some companies who participated in this research were measuring the number of maintenance items, the number overdue, and the number of times actions had been rescheduled. Incentives were then based on rate of completed maintenance that had not been rescheduled. This addresses a known unintended consequence in a straightforward way.

While most of our informants believed that the risk of unintended consequences was minimal, others were more wary. Consider the emphasis placed on minimising injury rates. This issue was mentioned as a current challenge. As a senior safety manager stated:

Does it drive the right behaviours? I am not so sure. I feel I am a guy of high integrity, and I'm a big advocate of very open reporting. Sometimes there is a very fine line between a medical treatment case and a first aid case. For sure I would say there are people on this planet who are fudging numbers, that is fair to say. In this company I have been concerned about a few assets and I have raised my concerns with headquarters. But generally I think we report things very openly.

A senior safety manager at another company explained that sometimes numbers were rationalised away:

The challenge we're having now in my experience is rationalising. 'I know it was a recordable injury' or 'I know it was a big spill' but then suddenly [we start discounting them because we had no control].... 'Yeah, one was some guy in the kitchen, one was a guy in the parking lot, I didn't have any at my unit!' Right? Then you start rationalising away. You can do the same with process safety. 'The spill was just barely an API Tier 1, and I really question how they calculated that number anyway, it was really a 2'.

This comment demonstrates an awareness of the potential for manipulation of occupational and process safety measurements. The company which reported this challenge was trying to combat this by focusing on severity and a qualitative assessment of overall safety performance, rather than the numbers themselves.

Reporting was the most common area in which unintended consequences were mentioned. One company explained that after including hazard reporting in their incentive arrangements, there was a significant increase in reports, including an unacceptable number of spurious reports. However, this company decided that as a result it would stop incentivising reporting. This is problematic. Ultimately in any system there will be reports that are of no use, but this is less of a problem than no reports at all. Other companies have addressed this challenge by moving to a metric that assesses the quality of the way reported incidents are addressed after reporting has been driven up to a sufficient level.

Consider, finally, the issue of close-outs, that is, the completion of action items stemming from an audit or an incident investigation. One of our interviewees commented on how incentivising close-outs can lead people to make do with superficial close-outs that are not in the long-term interests of a business. He explains:

I don't think we'd have anyone in a leadership role saying 'I can close all these out just by doing this' … If you sit down with any leader in the organisation and say 'what is the right thing to do' 99.9% of the time they come up with the same answer. Therefore I think you've really got to watch out where the head winds are so strong that it is causing people to take short cuts. It's not going to be 'I want to do it', it is going to be 'the only way I think I can do this here is…', 'I don't have the time' or 'I don't have the resources'.

In summary, while most of our respondents denied that unintended consequences might flow from incentivising certain indicators, others were well aware of this problem and provided specific evidence.

Is safety a special case?

One of the interesting findings that emerged in our interviews was the view that safety was a special case when it came to motivation. Either safety *did not need to* be included in incentive arrangements because senior managers were motivated by their own values in this matter, or alternatively, safety *should not* be included in incentive arrangements for various reasons. One of our interviewees provided an interesting argument to this effect.

> While there is a view … that the incentive scheme has a huge impact on people's behaviour, I would have to say in the safety space it doesn't play a big role. I say that because it is interesting to look at the behaviour of the organisation around injury rates. We have moved from a place where injury rates were 80% of the [safety component of the] bonus to 20%. And yet that hasn't changed the behaviour of the leaders – they are still very focused on injury rates.

Safety incentives are not necessary, we were told, because the values that managers personally hold make them concerned about the safety of their employees. Indeed, safety was a moral imperative that transcended financial considerations. It is necessary for us to examine this claim closely, as it appeared to influence the design of incentive arrangements and to legitimate their disproportionate focus on issues other than safety.

At one company, when asked about the perceived impact of incentive arrangements on senior executives, our informant commented that while the CEO was cognisant of the financial bonus, it was not what drives his decisions when it comes to safety. Rather, he has a "fundamental care and desire" to attend to it. Another commented: "I don't think financial rewards alone can drive you. The [safety] performance that we aspire to, I don't think it's something that has got to be bought ... It is part of the fabric of who we are." Both of these senior managers are claiming that safety motivations are independent of whatever incentive systems companies may create.

A challenge of researching an issue such as motivation by asking people what motivates them is judging when to take their answers at face value. Most managers will say that they do what they do because it is the right thing. As researchers, however, we also infer other motives from our observations and appreciations of the broader context. These may not be necessarily contradictory, but can lead us to conclude that a situation is more complex than is recognised by our interviewees. This belief that values can provide a sufficient motivation for safety needs to be treated with scepticism. The work environment creates immense pressure to achieve cost and schedule objectives. These may not necessarily negate safe decisions, but they can cloud focus. Equally, motivations are diverse and interconnected. Let us give an example to demonstrate this.

At one company, we spoke with three senior managers who were all keen to tell us that safety was a value in their company. Furthermore, they said, all staff were "hired and fired" on the basis of this value. This claim was repeated throughout the interview. We might interpret it to mean that, indeed, safety was in some way special, but this would be missing the reality of what is going on here. By their own admission, the motivation for individuals to give priority to safety does not reside in the value systems of the individuals concerned but in the threat of job loss. Of course it is the CEO who sets the scene for all of this. To deal properly with this issue we would need to examine the motives of CEOs, which we have not been able to do here. But it would be foolish to assume that their attention to safety is purely a function of their own value system.

It was also claimed that safety is special because it is a 'core' part of the role. Moreover, because it is core it need not be incentivised. A senior manager at one company captured this perspective: "It is an expectation of the business that we do our work safely. To be quite truthful, it would make no difference if I was being incentivised on that or not." This argument was sometimes given as a reason for not incentivising reporting. It was seen as a "part of the role". At one company the practice of "safety stewardship" – a core part of its safety assurance strategy – was not incentivised, for the same reason: "You can pay too much. Sometimes not so special performance gets rewarded. And there is always this fine line between what's really my job and what shall I get additional payment for."

There is a logical problem in all of this. If attending to safety is core to a role, so too is the achievement of business objectives. It follows that either both must be incentivised, or neither.

Another argument we encountered ran along the following lines. We cannot leave safety to the values of individuals. But nor can we expect financial incentives to provide the necessary motivation. Rather, financial incentives should be seen as *symbolising* company concern. Here is one expression of this view:

> This is a very live debate in our organisation. At one end of the spectrum safety is a value so it shouldn't be incentivised at all. At the other end of the spectrum there is a view that says incentives work in terms of making sure that people are focused. My view is that it is probably more symbolic than behaviour driving, in that if an organisation like ours that has high standards and expectations around safety didn't have something about safety in its incentive scheme it would look incongruous. You really are designing it more around making sure it has its place at the table rather than worrying about whether the percentage is big enough.

The senior safety manager who made this statement is claiming that incentives are symbolic, or that they broadly highlight priorities for the company. The precise metrics and their weightings do not matter. This is in many ways in keeping with our findings about the motivational capacity of performance agreements and evaluation processes not only for safety, but *all* metrics. If so, there is no need to treat safety as a special case.

The positioning of safety as a value can be interpreted as a way of managing the perceived conflict between business objectives and safety performance. One response to this conflict is to separate particular objectives into different realms: in the present case, safety is a matter of the values and moral commitments of individuals, while business objectives are not. According to philosopher Thomas Porter, this separation promises to structure our social worlds so that value tensions do not arise. There is, however, an important qualification. "Restructuring our social world so as to alleviate a tension in the principles guiding our actions is a worthy aim. But restructuring our social world so as to shirk some of their demands is not."[2] The cynic could reasonably suggest that positioning safety as special, as a value, may be a way for companies to escape hard questions about the low weighting given to safety in incentives.

To conclude this section we recount a story told to us by one of our interviewees about what can happen when delivery on schedule is incentivised, but safety, in particular process safety, is left to individual values. A major international oil company had commissioned the construction of a vessel that was to be anchored above a well on the sea floor so as to produce oil and gas. First oil was scheduled for 31 December and the timing was tight. The piping on the vessel needed to be tested in the shipyard before departure, to ensure it was gas tight. The test gas was supposed to be 99% nitrogen and 1% helium. Helium is much lighter than nitrogen and it is the helium that will escape if indeed there are leaks. The oil company recognised that there were problems with the piping and decided to do the leak test using nitrogen alone, which undermined completely the value of the test. Our interviewee, a contractor, became aware of what had happened and raised his concern. He was ignored. He decided that safety had been compromised to such an extent that he could not continue working on the vessel, and resigned. The vessel set sail from the dockyard on schedule, connected to the well and began producing oil by 31 December. As a result, all concerned had qualified for large bonuses. Four days later the vessel was forced to shut down because of leaks that took five months to rectify. As a contractor, our interviewee had not been eligible for a bonus and he was therefore, he said, not under the same financial pressure to compromise safety as was the case for the oil company managers.

Conclusion

This chapter has focused on the real experience of managers with performance agreements, performance evaluations and bonuses.

The findings seem remarkably inconsistent with company assumptions about how these things operate. In the first place performance agreements often do little more than restate the work objectives for the year and fail to provide clear criteria by which people will be evaluated. Secondly, the evaluation process is not tightly tied to the content of agreements, anyway. In fact it may only be in performance interviews that people get a clear indication of the real priorities of their supervisor. Third, the ranking process is counterproductive leaving most people without the positive feedback they seek. Most end up feeling that they have been damned with faint praise. Fourth, for most people money is not a direct motivator. Its significance lies in the signal it provides as to the way in which the supervisor has evaluated their performance. Fifth, arguments that safety can or should lie outside the incentive system lack force. Finally, we are not certain of the extent to which these conclusions apply to CEOs, who were not targeted for interview in this study.

A postscript

While we were carrying out this research, one of our companies, company B, became so disenchanted with the bonus system it was operating that it decided to radically alter it. We report on this innovation here because our findings suggest that it was a step in the right direction.

The particular catalyst was the discontent being generated by the rating system that classified people into four categories, the modal category being – "effective". People felt disillusioned by the fact that no matter how hard they worked, most of them would be told that they were merely effective. In addition, the concern about ratings was distorting the conversations that senior managers needed to have with their subordinates. These end-of-year performance conversations were perhaps the most important conversations that leaders had with their subordinates during the year, and yet their potential was not being realised.

On the other hand, financial bonuses were part of the remuneration landscape, and company B wanted to retain this mechanism for recognising performance. It therefore decided to abandon the system of categories and corresponding labels, while retaining the performance evaluation conversation and the financial reward. However, in order to be able to pay a performance-related bonus, it still needed to be able to differentiate people according to performance. This meant there was still a need for some kind of numerical rating. Accordingly, managers were asked to rate their direct reports on a scale of 1 to 100. They were told that most employees could be expected to be good performers and that they should typically be rated between 60 and 85. Managers were required to use a spread of ratings and the expectation was that there might be a few rated higher than 85 and possibly some lower than 60. Each manager might have perhaps a dozen people to rank and with so few people in the ranking pool there could be no expectation that any particular distribution be achieved. This numerical rating was then used to determine the actual bonus paid, as follows. For each employee the company determined a maximum available bonus, based in part on the company performance for the year and also on the salary level of the individual. The actual bonus paid was a percentage of this maximum available bonus, the percentage being the individual's personal score out of 100.[3] Individuals were told what their bonus was, but not the individual score that had been used to arrive at this amount. Finally, bonuses were treated as confidential, so it was not easy for people to compare their awards with others.

It is important to consider how this worked in practice. Mostly, leaders felt that the new system worked: it gave them the freedom to say what they needed to say. However, some leaders were not skilled at giving feedback, and continued using the old categories to express their evaluations. In other words, removing the categories did not automatically solve the issue. Company B therefore plans to conduct further training in how to engage in performance conversations.

Another problem was that employees were more concerned than usual with how the bonus was determined. In principle, it was possible for people to work out, based on the award they had received, just what score they had been given by their managers. In this way they could determine how they had been ranked relative to others in the pool. However, few people understood how to do this calculation, leaving them uncertain about the justification for the particular bonus they had received.

Although it was not easy to compare one's result with others, there was another type of comparison that *was* possible – to compare this year's bonus with last. The company's current year performance was worse than the previous year's and this meant all bonuses were down on the previous year. It was this that generated the most comment, in fact a groundswell of disappointment. This is revealing. It shows that people did not appreciate or accept that their bonuses might be affected by company performance, and they regarded the lower bonus they had received as somehow unfair. Moreover, if the variation in bonus from year to year was perceived to be a result of variations in company performance, rather than a reflection of one's own efforts, this would certainly have undermined the presumed incentive effect of the annual bonus system.

Perhaps the most disappointing aspect of the new scheme was that it continued to focus employee attention on the bonus and the way it was calculated, rather than on the conversations that managers had with their direct reports about things they had done well and ways in which they could improve.

For all these reasons Company B concluded that they would have done better to abandon bonuses altogether, so that the performance conversations were not contaminated by extraneous concerns. Whether this will be the next innovation is yet to be seen.

One final change has still to be mentioned. Company B had concluded that performance agreements listed far too many items. This made it impossible for people to know just where to focus attention. Under the new system, therefore, managers needed to specify only two or three objectives against which they would evaluate their subordinates. This was described as "decluttering" agreements. If readers refer to the typical performance agreement in Appendix 2, they will quickly see how necessary such decluttering can be.

In all, it is clear that Company B's experience is powerful confirmation of the conclusions we arrived at earlier in this chapter.

Notes

1 Doran, 1981.
2 Porter, 2009, p. 193.
3 There was a set budget available to each manager that constrained their ratings within certain bounds. There would have been a relationship between the set budget and the available bonuses, but we did not investigate this.

Chapter 7
Indicators of Major Hazard Risk

We demonstrated in Chapter 5 that process safety is largely invisible in the group component of bonuses. This is not to say that companies have been ignoring major hazard risk. Several in our sample had well developed process safety scorecards – sets of indicators – which they were using to good effect. For example, sets of process safety indicators appeared in the performance agreements of some of the managers we interviewed. As far as we can see, what has driven this development more than anything is the very high profile of the Texas City and Gulf of Mexico disasters, and a realisation that other companies may be just as much at risk as was BP. In particular there has been an increased understanding of the distinction between personal and process safety and the need for special focus on the latter. All this is perhaps best illustrated by the company whose relevant group safety indicator for bonus purposes was "TRCF+". As we noted in Chapter 5, process safety performance did not contribute to this indicator in a readily identifiable way, which meant that its impact on the group component of the bonus was not discernible. Nevertheless, process safety has had a much higher profile in this company over the last few years, a direct outcome, we were told, of the Texas City disaster.

In this chapter we would like to examine some of the ways companies have gone about developing indicators of major hazard risk in this post- Texas City environment. We shall also identify some of the problems we observed and suggest ways some of these indicators might be formally introduced into bonus systems.

Scorecards

We begin with the idea of process safety scorecards. Some we saw were simply lists of relevant indicators, but others were modelled explicitly or implicitly on the API 754 process safety triangle described in Chapter 3 (Figure 3.3). Figure 7.1 is an example from one of our companies, Company C.

This scorecard is a considered attempt to identify indicators related to each of the four tiers of API 754. The card extends to the right of what is shown here and contains columns for each of the company's assets and businesses. The boxes are colour-coded depending on whether the score is on target or better (green), falling short of target by less than 10% (amber), or falling short of target by more than 10% (red). Given that the full array consists of some hundreds of boxes, the traffic light system provides a useful visual display of the state of asset integrity in the global corporation.

Company C's attitude to red was ambivalent, as will become clear shortly. In the scorecard for the year we saw, about 10% of the boxes were red. The scorecard was accompanied by detailed text for each red box, explaining why it was red and whether there were acceptable reasons for the failure to meet the target. For example, several offshore platforms were behind on safety critical maintenance. This was because there were safety restrictions on the number of people and hence the number of maintenance workers allowed to be on platforms at any one time. Various mitigations were in place or being planned. Thus there was no automatic assumption that red was bad and needed to be eliminated at all costs, but rather that red was a flag that meant that the situation required extra scrutiny. The need for this more careful response to red is something the company had learnt from bitter experience, in relation to a similar personal safety scorecard. The previous year it had managed to convert all the reds to green by the year's end, but this had failed to result in any improvement in the TRCF. It concluded that its effort to convert red to green may have been too superficial, aimed simply at making the scorecard look better, without bringing about real change in the risk level. This is a striking example of the unintended consequences of placing too much emphasis on

	Indicator	Asset or Business			
		A	B	C	Etc.......
Tier 1	Loss of Primary Containment				
Tier 2	No. of LOPCs with greater consequence				
	No. of LOPCs with lesser consequence				
	No. of other LOPCs				
Tier 3	No. of demands on safety system/plant trips				
	% safety critical inspection results outside acceptable limits				
Challenge to safety system	% of safety system failure rate				
	% incidents with human error as causal factor				
	% of alarms outside target				
Operation management	No. of alarm floods				
	% of control loops on manual				
	No. of safety system overrides				
Maintenance	No. of safety critical work orders past completion date				
Tier 4	% emergency work				
	% of wells with safety critical maintenance past completion date				
Well integrity	% of wells with continuous annulus pressure problems				
	% of wells operating with safety critical elements under dispensation				
	% progress in asset integrity improvement plan implementation				
Process management	% of company standards under dispensation				
	% safety critical elements without performance standards				
	Compliance (% green)				

Figure 7.1   A process safety scorecard – indicators by asset or business

getting indicators to look good. Indeed, as we have previously noted, one major oil and gas company has a slogan – "challenge the green and embrace the red" – meaning: be suspicious of whether the green is really as good as it looks and treat the red as an indication of where attention needs to be focused, rather than as something to be converted to green as quickly as possible.

On the other hand – and this is why we used the word ambivalent at the beginning of the previous paragraph – company C had also chosen to summarise the situation for each asset along the bottom line of the scorecard. The summary measure was "% green", that is, per cent of the boxes in the column above that were green. Moreover it had set a target of 92% green for each asset. Finally, assets were summed along the bottom row to give an overall indicator of compliance. Again the target was 92%. From this point of view, green was good and red was bad. The target of 92% green was designed to drive improvement, but as we have already noted, such targets, where they are made to matter, can often result in behaviour that converts red to green, without any corresponding improvement in safety. It is perhaps just as well, therefore, that this measure had a very low profile in the company's bonus calculations. (In fact it was a single indicator amongst many in a general HSE scorecard, which itself counted for 15% of the overall group scorecard.)

Let us reflect a little further on why it is so problematic to include summary indicators of process safety scorecards in bonus arrangements. Such indicators are pure numbers. The context that has generated them has disappeared from view. They convey an evaluation divorced from what it is that is being evaluated. For this reason people at the top of the corporation, including remuneration committees, will be inclined to focus on how the number compares with whatever target has been set, not on why the number is as it is.

Although it seems neither feasible nor desirable to incorporate process safety scorecards, per se, into bonus arrangements, some of their constituent indicators may well be candidates for inclusion. A company may decide that it is appropriate in its circumstances to give priority to particular indicators by including them in bonuses. So for example, as we noted in Chapter 5, one of our companies had incorporated into its group multiplier: Tier 1 losses of containment, and a measure of the timeliness of safety critical maintenance. The beauty of this latter indicator is that it encourages people at the top of the corporation to ask questions about why safety critical maintenance is behind schedule, if indeed it is. Moreover, suppose that the answer relates to staffing limitations on offshore platforms. For people a few levels below the CEO in the hierarchy, this may seem like an insuperable problem, which can be mitigated but not resolved. From the CEO's perspective, however, the solution may be to stop production until the backlog is cleared, or to increase the capacity of platforms by improving evacuation options. Both of these may have financial implications that only people at the top of the corporate hierarchy could countenance. Having such an indicator in the company scorecard automatically elevates an issue like this to a level where something effective can be done about it. In particular, this may lead to the inclusion of relevant action items as qualitative elements in the performance agreements of the top people. We have already mentioned one example of this: "getting safety critical maintenance back on track".

We were surprised to find that two other indicators that quite commonly appear in process safety scorecards did not appear in any bonus calculations. They were:

- Number of approved deviations from approved engineering technical practices; and
- Number of safety system defeats (by-passes) that are currently in place.

We believe that companies should give consideration to including these items in corporate scorecards, as a way of drawing the necessary attention to them. Such indicators need not be in the scorecard indefinitely. Highlighting a problem in this way may lead to a fundamental solution, which enables the focus to shift to other indicators requiring a special attention.

Of course, as soon as any such indicators are included, one can anticipate attempts to manipulate the data. One company we have worked with included in its bonus arrangements a goal of zero overdue action items (items arising out of audits, incident reports etc). It then discovered that sites were achieving this target by *rescheduling* action items that had not been actioned, so that they no longer counted as overdue, which of course defeated the purpose. This is an example of the kind of unintended consequence that can flow from making indicators matter. As soon as a measure matters, attempts will be made to manage the measure itself rather than changing the behaviour that is being measured.

In view of this, the results for indicators that contribute directly to bonuses should be accompanied by a certification that the figures have been audited or carefully scrutinised by someone in authority. In particular, a sample of events that have been counted in, as well as a sample of those counted out, should be reanalysed by independent analysts to determine the reliability of the count.

Process safety scorecards are always works in progress and change from year to year as companies identify deficiencies and corresponding improvements. It is not surprising therefore that the process safety scorecard reproduced in Figure 7.1 is by no means perfect. One of the defects that stands out to us is the indicator "% incidents with human error as a causal factor". The fact is that undesired incidents nearly always involve human errors, if not by people at the front line, then errors by people making the many technical and managerial decisions in the lead up to incident. Moreover, as is often said but can never be said often enough, human error is the starting point for an investigation, not a conclusion. We will not labour the point here, but see Reason (1997).

Perhaps more importantly, the scorecard in Figure 7.1 is missing any measures of compliance with procedures. Process safety accidents are frequently triggered by failure to comply with procedures, such as start-up procedures, or permit-to-work procedures. It is far more difficult to assess compliance with such procedures than it is to assess compliance with some of the life-saving rules designed to prevent individual fatalities, such as driving rules. But process safety depends on procedural compliance just as much as personal safety depends on compliance with life-saving rules. Companies therefore need to audit procedural compliance as rigorously as possible and include some measures of this in their process safety scorecards.

Scorecards can be developed at many levels in large organisations, including at the site level. A processing plant where we did some interviewing had developed these ideas in an interesting way. According to its manager, everyone at the plant had been converted to "triangle" thinking: "We've been on a bit of a triangle journey and now it's got to the point at this site where it's not cool if you don't have your own triangle." The site had four main triangles, one each for personal safety, process safety, maintenance and reliability. Each was modelled on API 754. According to the manager, weekly leadership team meetings are spent standing in front of triangles. The numbers in the triangles are regularly updated, but more than that, the very categories themselves are continually discussed and modified to better reflect the reality of process risk at the refinery. These triangles, then, are constantly evolving.

Interestingly, many of the indicators in these triangles were specifically included in the site manager's personal performance agreement. Moreover that agreement included target numbers against which actual performance could be judged. According to the site manager, his results in relation to these targets had a significant influence on how he was evaluated in his annual performance review. So, although his company had no numerical measures of process safety in the company scorecard, *site* level numerical measures had been incorporated in the performance agreements of site managers in a systematic and influential way.

A contrary case

In contrast to the companies discussed above, some in our sample seemed to have learnt very little from the Texas City and Gulf of Mexico accidents. Particularly in companies for which pipeline operation was a major part of their business there was little awareness of the distinction between personal and process safety and no attempt to develop indicators of how well major hazard risk was being managed.

Company D (not a pipeline company) demonstrates some of the variability we encountered. It was still groping towards a clearer understanding of process safety, despite some very significant process incidents.

Until very recently the Group HSE measures relevant for bonus purposes were:

- TRCF;
- fatalities;
- sustainability;
- distribution incidents – category 2+.

Sustainability referred to greenhouse gas emissions and water consumption, reflecting the company's sensitivity to environmental issues. As for the fourth item in the above list, Company D provided the following definition.

> Distribution incidents are incidents occurring during transportation beyond the processing site. They are classified as category 2 if they meet certain threshold criteria: for instance, if they involve injury or damage, or if they involve a loss of containment that causes injury or damage or concern in the surrounding community or attracts local media attention.

There are several features of this definition that are worth remarking on. First, it is concerned with events off-site only. It seems not to matter how many losses of containment occur on site, provided the material released does not move off site. This is not an indicator, then, that tells us anything about the possibility of catastrophic on-site events. Second, the indicator is not restricted to losses of containment, but includes any transportation accident that causes injury. In other words there is no distinction being made here between personal and process safety. Third, the reference to community concerns and media attention suggests that damage to reputation may be what Company D is most worried about. This would account for the fact that incidents are regarded more seriously if they occur off site rather than on site. Perhaps not surprisingly, Company D was not managing process safety effectively, something it finally realised when it had some significant on-site releases that affected residents off site. This led among other things to a re-evaluation of the distribution incident measure and its replacement by an indicator known as "process excursions". This was defined as "a serious loss of containment that occurs on or offsite", with serious, in turn, defined.

This new indicator is clearly a step in the right direction. But to describe such incidents as process excursions reveals a considerable lack of understanding. Process excursion is almost universally understood to refer to situations in which a process parameter such as temperature or pressure is beyond normal operating limits. Process excursions do not necessarily or normally lead to a loss of containment. Process excursions, so understood, are an important indicator of process safety, but to call a loss of containment a process excursion, as Company D is now doing, simply confuses the issue. One would have to say that Company D, a multinational operating in several countries, still has not absorbed the lessons of Texas City.

There is one other piece of evidence of company D's failure in this respect. As a general rule of thumb companies that take process safety seriously have a process safety expert on the executive committee, that is, reporting directly to the CEO. BP had no such person prior to Texas City. Following the Gulf of Mexico accident it created just such a position. At the time of our interviews, Company D had no process safety expert reporting directly to the CEO.

Precursor events

The earlier discussion of process safety scorecards begs an important question. Why use scorecards at all? Why not use more direct indicators of major hazard risk? After all, with certain qualifications discussed in Chapter 2, injury rates and fatality rates are reasonable indicators of how well personal safety risks are being controlled. Can we not find a similar indicator that will tell us fairly directly how well major hazard risk is being managed?

The idea of a single indicator of process safety has been something of a holy grail. It informs, for example, the thinking that lies behind the API 754 process safety triangle described in Chapter 3 (Figure 3.3). The theory, as explained there, is that in the facilities covered by API 754, such as refineries and gas processing plants, a loss of containment, is a precursor to a major accident. It denotes a situation of elevated risk of explosion and fire and hence, the fewer such losses of containment, the better. However, many companies have somewhat uncritically accepted the idea that loss of containment is the long sought-after single indicator, and in so doing have generalised inappropriately. For example, the scorecard for Company C (Figure 7.1) treats LOC as the most important single indicator across all its businesses. Given that most of these businesses are engaged in oil and gas extraction and processing, this may be appropriate, since reducing the number of such loses reduces the risk of their most catastrophic scenarios. However, one of Company C's businesses is shipping. Here, the most catastrophic scenario is probably collision or grounding, resulting in a major loss of containment. Ordinary loss of containment events are not precursors to such a scenario and reducing the number of ordinary LOCs does not decrease the risk of catastrophic collision or grounding. Relevant precursors to such disasters are occasions on which vessels are on a collision course, or simply off course. So, in treating loss of containment as its most significant risk indicator for its shipping business, Company C is systematically misleading itself. In the scorecard for the year we saw, the shipping business experienced no Tier 1 or Tier 2 losses of containment, yet it might well have been at significant risk of experiencing a catastrophic collision. The company's risk indicators were not geared to the possibility.[1]

Offshore oil and gas

Let us generalise the preceding discussion. Industries need to identify all the major accident events they wish to prevent and then identify precursors relevant to each. Each such precursor event is a situation of elevated risk and hence reducing their number reduces the risk of the major accident event in question.

An excellent example of this approach is to be found in the Norwegian offshore oil and gas industry, led by the regulator, the Norwegian Petroleum Safety Authority. The Authority has identified a series of events that are the precursors to different kinds of major accidents.[2] The most important are identified in Figure 7.2.

These are all indicators of increased risk with respect to some major accident scenario of concern. The scenarios are diverse, including major fires on production platforms, blowouts on

1. Non-ignited hydrocarbon leaks

2. Ignited hydrocarbon leaks

3. Well kicks/loss of well control

4. Fire/explosion in other areas, flammable liquids

5. Vessel on collision course

6. Drifting object

7. Collision with field-related vessel/installation/shuttle tanker

8. Structural damage to platform/stability/anchoring/positioning failure

9. Leaking from sub-sea production systems/pipelines/risers/flowlines/loading buoys/loading hoses

10. Damage to sub-sea production equipment/pipelines/diving equipment caused by fishing gear

11. Evacuation (precautionary/emergency evacuation)

12. Helicopter crash/emergency landing on/near installation

Figure 7.2   Precursor events used in the Norwegian petroleum industry

drilling rigs, helicopter crashes and various kinds of collisions at sea. Clearly there is no single indicator relevant to all these events, nor is there a one-to-one relationship between the indicators listed in Figure 7.2 and the scenarios of concern. Together, however, these indicators provide a comprehensive risk picture. This is a state-sponsored initiative, but it is clear that every company operating in a hazardous industry would do well to follow this model. Specifically, every company should develop a suit of indicators which measure the frequency of the immediate precursors to each distinct major accident event that the company could potentially experience.

This discussion can be put into a broader context by comparing Figures 7.2 and 7.1. Figure 7.1 identifies a single precursor event – a loss of containment – and then lists a series of "lead" indicators that, if managed properly, will reduce the risk of a loss of containment event, which in turn reduces the risk of fire/ explosion. Figure 7.2, on the other hand, identifies a series of precursor events, precursors to different kinds of major accident events. It does not identify ways in which the frequency of such precursor events can be reduced, so as to reduce the risk of that particular major accident event. To complete the picture sketched in Figure 7.2, for each of those precursor events a series of lead indicators needs to be identified and managed.

High pressure gas pipelines

In contrast to the Norwegian offshore oil and gas industry, the high pressure gas pipeline operators in our sample had not yet understood the importance of identifying relevant precursors. We asked the managing director of one such company: "What is the most worrying scenario for your company?" He was quite clear that the potential for a full-bore rupture of a high pressure gas pipeline was top of his list. If this happened on a high pressure line, the impact zone of any explosion and fire would be around a kilometre. In populated locations, this could be a disaster with loss of life on a catastrophic scale.

The industry view is that the most likely cause of such an event is third-party interference: encroachment on an easement with plant such as an excavator, augur or horizontal borer that results in a pipeline rupture. As in the shipping case, the most catastrophic scenario of a pipeline rupture results in a major loss of containment. But, just as in the case of shipping, smaller loss of containment events are not precursors to this most catastrophic possibility. Instead, an indicator more sensitive to the management of this major risk would be unauthorised encroachments. If we follow this scenario a step further, we note that a small proportion of unauthorised encroachments result in a strike on the pipeline (but not necessarily penetration). So a second indicator of the risk of a pipeline rupture is number of pipeline strikes. However this indicator is likely to be less useful, since the number of strikes may be very small. If so, it is better to manage and seek to drive down the number of encroachments rather than the number of strikes. Yet among the companies we interviewed that were operating high pressure gas pipelines, not one was using either of these metrics as indicators of major hazard risk, let alone including them in bonus arrangements. The best of them were using loss of containment as their principle indicator of major hazard risk, but had failed to realise that this gave them no indication of how well they were managing the risk of third-party damage leading to a catastrophic release.

Another disaster scenario for high pressure gas pipelines is an original manufacturing defect that many years later may lead to full-bore rupture. Other defects include cracks, dents, and pits and may be caused by corrosion or fatigue. Depending on the exact cause, defects of these kinds can be identified by various inspection methods, such as external direct assessment, or by traversing the line internally with a robotic device known as an intelligent pig. Appropriate safety metrics, then, might be the number of inspections and tests conducted in a given period and actions to address identified faults. Only one of our companies operating a pipeline business explicitly included such measures.

It is of concern that several of our companies (both in the pipeline and other hazardous industries) did not include metrics thoughtfully targeted towards their worst case scenarios. Our interviewees in these companies typically knew what needed to be measured to give them a sense of the overall safety of an asset. However, these measures had no visibility in their companies and had certainly not found their way into incentive schemes.

Conclusion

As we have demonstrated in this chapter, some companies have made considerable progress in developing indicators of major hazard risk. This progress is based on the thinking implicit in the API 754 process safety triangle and the UK Health and Safety Executive barrier performance indicator model, both discussed in Chapter 3.

However many companies have failed to recognise that API 754 was devised for the refinery and petrochemical industries, which means that it is not necessarily appropriate for other major hazard industries, and in particular is not necessarily appropriate for other segments of the oil and gas industry. Each industry and industry segment must work out for itself what are its most feared major accident events and identify precursor events that can be counted and converted into an indicator of risk to be driven downwards.

But assuming this issue has been dealt with, the challenge is to find ways to incorporate indicators of major hazard risk into bonus systems. Losses of containment, though not ideal, may be a suitable indicator for inclusion in top-level scorecards for many companies. Moreover, depending on the circumstances of a company, one or two specific indicators measuring the effectiveness of controls might be included, as a way of directing the attention of top-level management to particular issues. We argue that traffic light summaries (e.g. % green) are not helpful because the number is so remote from any meaningful referents and because such summaries are particularly likely to encourage efforts to manage the measure.

One of the most effective strategies is the inclusion of *site* level process safety indicators into the performance agreements of *site* managers. This ensures that the indicators are indeed relevant to the operations at hand. The only drawback to this strategy is that the performance evaluation process is not tightly linked to what is in performance agreements and it is only if supervisors choose to emphasise these indicators in their evaluations that they will be perceived as mattering. This in turn depends on a commitment from the people at the top to ensure that process safety indicators are given prominence in the performance evaluation process.

Notes

1   Here is another disastrous scenario. When hydrocarbon tankers are loading and unloading, there is a possibility that the atmosphere inside the tank will not be properly inerted and will become explosive. Static electricity could then spark an explosion with potentially major loss of life. We are unaware of cases where this has happened, but this is a well-known risk. We have not seen any attempt to identify precursors to such events and include them in scorecards.
2   Petroleum Safety Authority Norway, *Trends in Risk Level*, Summary Report 2009.

Chapter 8
Summary, Reflections, Suggestions

Reports following the BP Texas City refinery disaster demonstrated how bonus arrangements had distracted attention from process or major hazard safety. These reports therefore proposed the restructuring of bonuses to include a focus on process safety. This is a deceptively simple idea. In fact it raises a number of questions, about how bonus systems are designed, what might be involved in focusing them more effectively on catastrophic risk and what it is that really motivates people in the corporate context anyway. This book seeks answers to these questions and concludes with some proposals of our own.

The first part of the book lays out the issues and indeed some of the answers to our questions, but we recognised that we could not answer our questions fully without a further empirical investigation. Accordingly we targeted a total of 11 companies in the resources sector, mainly multinationals. We studied documents and interviewed a large number of senior managers in these companies. The second part of this book is a description of what we found. In this final chapter we sum up these findings, reflect on them, and make some suggestions.

Long-term bonuses

Bonuses are an extremely important component of the remuneration paid to the top people in corporations we investigated, potentially worth several times more than their fixed salary. In the case of one CEO, this variable component of remuneration could be up to eight or nine times the fixed component! The stakes in other words are very high for people at the top.

There were two main types of bonus identified in our research – long term and short term. Long-term bonuses were generally paid to a restricted group of top people and were substantially larger than short-term bonuses. It is crucially important, therefore, to understand how they work and to identify their potential impact on safety. For simplicity we focus this discussion on CEOs. As we explained in detail in Chapter 4, long-term bonuses are carefully designed to encourage CEOs to maximise shareholder return. They are paid in company shares and are awarded annually on the basis of company performance, largely economic performance. However these awards are not paid in the year in which they are earned. Rather, they vest, meaning they are actually paid, some years later, typically three. If awards were simply deferred in this way, this would, in itself, provide CEOs with an incentive to maximise the share price in the interim, since they themselves would be beneficiaries of any price rise. But remuneration committees have designed a much more powerful incentive. The shares only vest if a company does well in comparison to its direct competitors. If the total shareholder return in the years prior to vesting is less than the median value for a carefully chosen competitor group, then the entire bonus due to vest that year is obliterated. In other words, it is not enough that the company does well and that shareholders are getting a good return on their investment. That return must be at least equal to that of the middle ranking company in the comparator group *for any bonus at all to be paid*. This puts enormous pressure on CEOs to do whatever is necessary to ensure that the company is up with, or better still, ahead of the pack.

Consider now the impact of such bonuses on safety. Most accidents have no discernible effect on profit, so long-term bonuses provide no incentive to reduce the number of accidents. On the contrary, long-term bonuses provide an incentive to maximise profit at the expense of safety, if need be.

There is one possible qualification to these general propositions. Major accidents, though rare from the point of view of any one company, can have a substantial effect on shareholder return. The argument put to us by our interviewees was that this meant that long-term bonuses *did* provide CEOs with an incentive to manage catastrophic hazards effectively. Moreover, if the vesting period was prolonged for up to ten years, we were told, this would encourage CEOs to take the long view and to recognise that it was indeed in their interest to manage carefully the risk of rare but catastrophic events.

However the specific structure of long-term incentives described above undermines this potential. It is true that a major accident could wipe out the CEO's long-term bonus. But such an accident is very unlikely to occur. On the other hand, spending more on the management of major hazards may well reduce profit sufficiently to put the company fractionally behind its competitors, again wiping out the long-term bonus. This is a much more likely scenario. From this point of view the rational CEO will choose to maximise current shareholder return, regardless of the long-term safety consequences. In other words, the particular bonus structure described here fails to provide CEOs with an incentive to manage carefully the risk of rare but catastrophic events.

We argued in Chapter 4 that the long-term bonus structure described above is not in the interests of many shareholders. It is of no particular concern to most shareholders whether the company falls just above or just below the median mark. What is of interest to many institutional shareholders, such as pension funds, is that companies manage catastrophic risk effectively. UK pension funds held substantial portfolios of BP shares and were severely affected by the drop in share value following the Gulf of Mexico disaster. It is worth noting that most people investing in pension funds opt for the balanced investment strategy, implying a preference for stability and security rather than riskier profit-maximising strategies. It is not in the interest of such people that the resource companies in which their pension contributions are invested should remunerate their CEOs as they do.

Some alternatives

There are other bonus structures that do not have the consequence discussed above. Suppose the long-term bonus was not subject to total wipe-out if total shareholder return fell slightly behind other comparator companies, but simply declined in some proportionate way. CEOs would then face a different choice. One option would be to spend whatever was necessary to manage catastrophic risk effectively, accepting that this may reduce their long-term bonus slightly. The other option would be to minimise such spending, thereby maximising shareholder return, but also increasing the risk of a major accident that might ultimately wipe out the entire bonus. This is a very different financial risk management decision for CEOs, which might well lead them to give greater priority to avoiding rare but catastrophic events.

Be that as it may, we would like to propose an alternative possibility, first suggested in the context of the financial industry (see Chapter 2). It is that a sizeable proportion of deferred bonuses be paid as cash directly into a trust fund – a bonus pool – that could be drawn on to pay for damages in the event of catastrophic incidents. People whose money was tied up in such a trust fund would therefore have a vested interest in minimising the risk of catastrophic incidents. Such a fund might operate as a unit trust.[1] Under such an arrangement, deferred remuneration would be used to buy additional units in the trust, at the prevailing price, and units could be sold when the period of deferral had expired. The value of the units at the time of sale would be augmented by any profit or interest paid to the trust between the purchase and sale of units, but diminished by any pay-outs by way of compensation to those affected following major accident events.

An important feature of this arrangement is that the value of units in a trust will not be affected by a company's financial performance. (The funds of the trust should not be invested in the company itself.) Rather the value of units will be much more directly affected by how well the firm is managing catastrophic safety and environmental risk. This is a highly desirable feature of the scheme from the present point of view.

Such a scheme need not be limited to long-term bonuses. As we noted in Chapter 4, short-term bonuses are deferred under certain conditions and such deferred bonuses could also be paid into the trust. Indeed it would make sense to defer bonuses in this way for substantial numbers of senior people.

A unit trust scheme of this nature would mobilise safety expertise in a novel way. Not only would it encourage senior people to give a high priority to the management of catastrophic risk in their own areas of responsibility, it would also encourage them to take an interest in what their peers were doing in other areas of the business, since a major accident in any part of the business would have an impact throughout the business. Moreover those who had recently retired from senior managerial and technical positions and were still awaiting final payouts from the bonus pool would have a vested interest in ensuring that their firm or business unit continued to manage catastrophic risk properly. Not only would they have an interest, they would also have time to devote to ensuring that their former colleagues were doing the right thing. They would be well aware of the issues and they would probably be able to exercise influence by means of direct phone calls. They would constitute what Jacobs describes as a "Jesuitical corps"[2] of trusted and expert people looking over the shoulders of decision makers and keen to provide advice. This is perhaps one of the most interesting and potentially far-reaching consequences of the unit trust proposal. It involves mobilising additional pairs of eyes and ears – expert eyes and ears – in an unprecedented way.[3]

The preceding discussion provides a set of ideas for constructing deferred bonus schemes that encourage a company focus on disaster prevention. There is no reason why such a scheme could not operate alongside a scheme designed to maximise shareholder return. CEOs would then have a real incentive to consider carefully how to balance profit maximisation and catastrophic risk reduction. It is unlikely that companies would voluntarily introduce such a scheme; it would probably need to be legislatively prescribed.

Health disasters

The focus in this book is on safety and environmental disasters. But before moving on it is worth speculating about the potential for deferred bonus schemes to prevent *health disasters*, such as the enormous toll of death and disease caused by asbestos and by tobacco. Both asbestos and tobacco companies have been faced with massive compensation payouts in recent years. In both cases there was evidence many decades previously of the damaging health effects of their products, which the companies ignored, indeed suppressed, as long as they could.

Suppose a bonus scheme with a seven- or ten-year deferment had been in place. What difference might this have made? At first glance, not much. When the evidence began to emerge that these products were dangerous, before World War II in the case of asbestos, court-awarded damages were still decades away. On the other hand, the decision makers at the time had no way of knowing how far into the future the first big compensation awards might be. Had a scheme been in place they would probably have begun to act on the assumption that their deferred benefits might indeed be affected. Eventually it might have proved difficult to attract good people to such jobs and these industries might have been wound back much sooner than they were.

Such a suggestion involves an extension of deferred bonus schemes, from industries that can experience sudden catastrophic events, to those where disasters develop over a long period of time. This raises the question of just which industries could be covered by such arrangements. In principle we would want them to operate in all industries where there is a potential for disaster. In practice, however, we would need to proceed incrementally. One possibility would be to extend these arrangements to any industry where insurers are unwilling to provide insurance because of the possibility of catastrophic losses. As early as 1918, some US insurers would not offer life insurance to asbestos workers,[4] and this could have been the trigger for governments to mandate deferred bonus and share option arrangements for companies dealing with asbestos.

Short-term bonuses

The second main type of bonus being paid in our sample of companies is the short term, that is, annual bonus. Short-term bonuses are paid to many more people at many more levels of management than is the case for long-term bonuses. The further down the hierarchy, the smaller the bonus is as a proportion of total remuneration, but for many senior managers, the annual bonus can equal or exceed the fixed salary. Moreover, short-term bonuses are generally paid in cash at the end of the year in which they are earned, although for very senior people a portion of this bonus may be deferred. Safety is systematically incorporated into the calculations for short-term bonuses, so in theory they have the potential to enhance performance, with respect to both personal and process safety. But before coming to conclusions about this we need to examine in more detail how they work.

Short-term bonuses in our sample of companies were generally determined by evaluating two types of performance – the performance of the group and the performance of the individual. The group was generally the global corporation, but it could also be one of the constituent businesses or assets or even a combination of these. These two components – group and individual – were combined in one of two ways: they were either added or multiplied together. Most companies were using the multiplicative approach. A feature of this approach is that a poor individual performance coupled with a poor group performance yields a very small bonus, while a superior individual result coupled with a superior group result yields a very large bonus.

Group performance

The group performance was generally evaluated using a series of quantitative measures, dealing with profit, production, injury rates and so on, for the whole group, generally the whole corporation. Targets were set for each of these measures and actual results compared with the target to provide a score for each of these indicators. These scores were weighted according to an agreed formula and added together. One of our interests was the weighting given to safety in these calculations. The figure was in the range of 10% to 30%. In other words, safety, in all its forms, was usually a relatively minor contributor to the group performance scorecard and hence to the group component of the bonus. One company had managed to prevent economic indicators from swamping the safety indicators by multiplying the results rather than adding them. This meant that in order to end up with a good group result, both safety and economic indicators needed to be good. Provided the computations don't become too complex, this would seem to be a sensible way of ensuring that safety remains an influential component of the group scorecard. Finally, and most significantly from our point of view, with few exceptions the safety measures referred mainly to personal safety. Process safety was largely invisible in these group scorecards.

Despite the effort that companies had put into developing scores and targets for purposes of measuring group performance, they all acknowledged that the ultimate evaluation was qualitative, relying on judgement, and not the mechanical application of a formula. This of course muddied the waters, making it ultimately impossible to be sure how much weight had really been given to safety.

We come now to one of the most paradoxical aspects of the whole short-term bonus system. The great majority of managers have no capacity to influence corporate results in any perceptible way. To use the language of our interviewees, they do not have line of sight to these figures. Only the people at the very top of the corporation can expect to influence the overall numbers by the decisions they take. For everyone else, the corporate results are beyond their control; if not in the lap of the gods, then certainly in the lap of someone far above them in the corporate hierarchy. It follows that these corporate results cannot serve to motivate their behaviour.

This problem is widely recognised and is sometimes used as an argument for using the results of constituent businesses or assets as the relevant group results for bonus purposes, rather than the results for the whole corporation. Despite this logic, the tendency has been to move in the reverse direction, by focusing on the global, corporate results, rather than more local group results. The purpose has been to convey the message that the global or corporate result is more important than the results of the constituent businesses. But whatever the symbolism, the insistence on using the global corporation as the relevant group for bonus calculations seriously undermines the motivational value of this aspect of the annual bonus. The group component of the annual bonus cannot serve to motivate people with respect to safety performance or anything else. It should be abandoned, unless a much clearer justification can be offered.

There is one important exception to what we have just said. CEOs and some of their direct reports do have the capacity to influence corporate outcomes and it therefore makes sense to include indicators of group performance in their bonus arrangements. We return to this below.

Individual performance

Individual performance was evaluated on the basis of a performance agreement that each person had with his or her supervisor. These agreements generally contained qualitative goals rather than quantitative targets. The situation was different for the heads of groups whose outputs could be measured, for instance, CEOs, business unit leaders and some asset managers. For such people, the performance agreements often specified group targets – production, profit, injury rates and so on – for which the group manager was held personally accountable. This made perfect sense, since those in charge of these various enterprises were indeed in a position to influence the outcomes.

For most people performance agreements consisted of lists of things they intended to accomplish during the year, designed to support the objectives of the group. There might also be some vaguely worded objectives such as "contribute to the cost reduction goals" of the corporation. Sometimes agreements included quantitative targets measuring some aspect of the individual's activity, such as doing a certain number of safety observations.

Safety was a relatively small component in these agreements, particularly for people who could not be held accountable for group rates. Sometimes an attempt was made to specify that safety accounted for a certain percentage of the overall individual performance evaluation. But where this was the case, the percentage was small, less than 25%.

Regardless of any such weightings, there was no way an individual's performance could be evaluated quantitatively. Instead a performance evaluation interview was conducted by the supervisor who made an overall, subjective assessment and ranked the individual in one of a small number of categories, such as: below expectations, meets expectations, exceeds expectations, or outstanding.

But these are not unconstrained evaluations. Companies often specify what proportions of people should fall into each of these categories, requiring in particular that the majority of people end up ranked as "meeting expectations". To facilitate this, peers are grouped into comparison pools, and supervisors of all the people in such a pool must agree among themselves on how their rankings need to be adjusted in order to conform to the required distribution. Beyond this, people may be fully rank-ordered and each assigned a unique score corresponding to that rank-ordering, for purposes of bonus calculations. Again, supervisors must debate with each other and agree on the rank-ordering of all the people in the comparison pool.

From the point of view of supervisors, this moderation process by which initial evaluations are adjusted is difficult and time-consuming, even distasteful. From the point of view of people evaluated in this way, the process is demoralising and disheartening for the majority, since they are essentially being told that there performance is average. No matter how hard they may try, most are condemned to be evaluated as satisfactory. In their minds, this amounts to being damned with faint praise.

Furthermore, for most people, particularly as one moves down the hierarchy, there is not a great deal of money at stake in these evaluations. What is really important to people is the evaluation itself and a sense of whether or not one is seen to be doing a good job, as opposed to an average or satisfactory job. Positive feedback is what people crave, yet this is exactly what the process denies to most of them.

It is curious that businesses have dug themselves into the hole we have described; remuneration consultants appear to have swept all before them. Fortunately, this disheartening evaluation process is easily avoided. One alternative, described in Chapter 2, is to make the performance review a prelude to being awarded an annual increment. The great majority of people could then be told that their performance is good and warrants the increment.[5] The performance evaluation would thus be transformed into a positive experience. Of course there will be some whose performance is not satisfactory who will be denied the increment. Furthermore, there is no need to fear that such a system would remove the incentive to do better. It is not only increments but also promotions that are at stake, so there is still plenty of material incentive to please the boss, quite apart from the psychological rewards. We can personally vouch for this as academics. We do not receive annual bonuses based on performance, but there is no shortage of incentives in our organisational environment to encourage us to give of our best. Finally, in the performance interviews that we advocate, people can not only be congratulated on their performance, but also coached as to how they can do better. These interviews can therefore perform all the functions of the performance evaluation interviews that companies are currently using.

Given that performance agreements are quite vague on what the priorities are, and given that supervisors in any case form an overall view of the individual's performance, the question arises as to the real determinants of that judgement. What we found is that evaluators sometimes pay scant attention to what is actually in the agreement and may even evaluate people on the basis of behaviour that is not specified in the agreement. Furthermore, supervisors may convey a very clear idea of what is most important by singling out certain matters for the hard conversations about where people need to do better. We were struck by the case of one refinery manager who was told that his injury rate was too high and that it was his responsibility to bring it down. This was, he said, a defining moment in his career, meaning, a moment in which he became a lot clearer about what his priorities needed to be. So while the performance evaluation process can be very effective in setting priorities, it is not at all clear from an examination of performance agreements themselves just what those priorities are. It is not surprising therefore that some of our interviewees seemed almost unaware of the contents of their performance agreement.

It would seem that performance reviews serve, among other things, to convey priorities of supervisors. Those priorities in turn may stem from *their* superiors and so on up the line. What then is the starting point? If we pursue the matter of injury rates for a moment, large companies are under considerable scrutiny with respect to injury rates and this naturally becomes an indicator in the performance agreement of CEOs. If it becomes a priority for the CEO it will become a priority for everyone in the chains of supervision below – "what interests my boss fascinates me". A culture is created in this way that does not rely on the performance agreements, as such, but rather on the evaluation interviews that emphasise the importance of this issue. This is possibly the best explanation for why it is that injury rates were such a focus of attention in the companies we studied – not so much because these injury rates appeared in performance agreements, but because supervisors emphasise this in their performance reviews. This was part of the culture of these organisations, set in place by the priorities of the CEO.

To summarise to this point, we believe that large resource sector companies should dispense with annual bonuses for most people. The group component of the bonuses serves no motivational purpose for people who can have no perceptible influence on group performance. However the group level indicators *can* be incorporated into the bonuses of CEOs and some of their direct reports. As for individual performance agreements, these can serve as a useful basis for performance reviews, but a process that evaluates most people as average or satisfactory is unnecessarily destructive of morale and should be abandoned. It could be replaced by a performance evaluation aimed at determining eligibility for a pay increment. The great majority of people would be judged eligible for the increment and hence for the great majority of people the performance review would be a rewarding experience. Where performance evaluations are based on quantitative data, as they can be for business unit leaders, it does make sense to evaluate and remunerate people accordingly. However this does not need to be done relative to others; it can be done simply on the basis of the business unit performance relative to targets.

We can expect that global and business unit leaders whose priorities are determined by numerical indicators and targets, will cascade these priorities to their subordinates. This may be in the form of subgroup indicators and targets, where relevant, but more commonly by specifying activities and projects that will support the aims of the business unit. Most importantly, leaders who are driven by certain performance indicators will create cultures in which these things are valued, regardless of whether they can be measured.

The role of process safety indicators

Some companies have done good work in developing sets of process safety indicators or scorecards. However there is a tendency to assume that loss of containment is the best precursor indicator in all contexts. This is not the case, and businesses at all levels need to think carefully about the major accidents from which they are most at risk and the relevant precursors to such accidents. They should then include indicators of the frequency of such events in their scorecards. For instance, loss of containment will not be the best indicator for drilling, shipping, aviation, road tanker operations and pipelines. More relevant to these activities may be the following: kicks, vessels on collision course, sudden braking, and third-party incursions or strikes, respectively. Once this context is set, companies can develop indicators of how well these risks are being managed. Quite complex scorecards can be built up in this way, at each organisational level – such as Group, business unit and asset or facility.

The crucial question, though, is the extent to which these scorecards should be incorporated into bonus arrangements. The only way that scorecards, as a whole, can be included is if they are summarised in some way, such as per cent green. However such summaries inevitably lose sight of the reality of process safety and almost inevitably give rise to attempts to massage the measure. Far better that particular indicators be included in the performance agreement of particular managers who have the capacity to influence them. For the CEO it may be that a single indicator measuring losses of containment is the most appropriate. To the extent that this indicator can be made to matter at the top of the company, we can expect top leadership will generate a culture of concern about process safety that will percolate downward. This will lead to a recognition that loss of containment may not be the most appropriate process safety indicator for some constituent businesses, such as drilling, and that the leaders of such businesses need to be incentivised differently. Other indicators relevant to process safety, such as the timeliness of safety-critical maintenance, can be included in the performance agreements of very senior people, in the expectation that this will encourage them to make the necessary resourcing decisions to ensure that, in this case, all safety-critical maintenance is up to date. We stress, however, that where such indicators are made to matter in this way, there is an inevitable incentive to manage the measure, rather than that which is being measured. Accordingly, all such results must be treated sceptically and subjected to independent verification. They should not be accepted unless someone in authority is prepared to vouch for them.

Including safety in performance agreements

We have argued that the actual content of performance agreements is not critical to the performance evaluation in many cases. This is not to say that it is irrelevant. Performance agreements set goals and influence to some extent the content of the conversation at the time of the performance review. It is therefore worth ensuring that safety, particularly in relation to catastrophic risk, is included as effectively as possible. We have noted above that numerical process safety indicators are relevant in some cases. Moreover we provided examples in Chapters 3 and 5 of innovative ideas tailored to the specific functional positions – finance managers, legal counsel, human resource managers and so on, which we shall not repeat here. There is one additional point worth mentioning. It is generally agreed that a good safety culture is one where incidents are reported and lessons are effectively learnt.[6] Performance agreements can readily be used to promote such a culture. Suppose the agreement included a requirement to "set up a system in your area that successfully encourages the reporting of bad news and ensures that these reports are responded to effectively". Such a system was described in Chapter 3 in which an operator was given a diamond award for reporting that the operating limits of a machine had been changed without going through the prescribed process. A requirement in a performance agreement for site or asset manager to set up such a system could well see the development of a variety of innovative ideas, and performance review conversations could help to improve these ideas in subsequent years. Finally, the policy adopted by Company A in Chapter 5 is worthy of consideration. Under that policy senior managers were required to certify personally that critical controls relevant to fatality risks at particular sites were in place and functioning.

Human motivation

Chapter 2 raised the question of whether financial bonuses are an effective way to motivate people in a corporate context. We reflect, finally, on this question. Pink and others argue that human motivation is complex, including not only a desire for money but also, a desire for status, job satisfaction, appreciation by others and fairness, among other things. Few would disagree with this. Pink takes this a step further by arguing that any attempt to motivate people in a corporate context by offering financial inducements misses all this complexity. However Pink's claim itself is missing the context in which bonuses, in particular annual bonuses, are paid. They are awarded in the context of a performance review in which people's performance is evaluated. Our conclusion is that this evaluation is considerably more important as a motivator than the money that may go with it. The best evidence for this claim is the fact that the evaluation process, which requires that people be distributed on roughly a normal curve, leaves the majority feeling that they have been damned with faint praise, rather than appreciated for their contribution.

Furthermore, most of our respondents denied that the financial aspect of bonuses was the main motivating factor. One senior executive explained that he was very well paid and did not need the additional money. But, he went on, "I like to be recognised for doing a great job ... and I like to be recognised for that in monetary terms." This comment captured much of the sentiment we encountered and the ambivalence about the nature of the incentive.

When it comes to long-term bonuses, the situation may be slightly different. Such bonuses can be in excess of 200% of salary and some of our respondents were of the view that when such vast amounts of money were involved this was likely to become a primary motivator. However it is possibly more relevant that the context in which long-term bonuses are awarded is different. There is no evaluation by a supervisor. Rather, it is the cold hard facts of the share market that determine whether a bonus is paid or not. This would seem to be a situation in which financial motivations will be paramount. But even here it is possible to point to certain psychological aspects. Given the way the bonuses are constructed, a CEO who gets his or her company into the top half of the distribution, and particularly into the top quartile, will be lauded by the Board, financial commentators and others. The existence of this psychological reward makes it difficult to be dogmatic about the precise way in which long-term bonuses motivate their recipients.

At the end of the day, therefore, this study does not significantly advance our understanding of human motivation, since we have not been able to disentangle the effects of financial from other forms of motivation in a definitive way. What we do demonstrate is that bonus systems tap multiple motives, and we use these insights to suggest how bonus arrangements can be made to operate more effectively, particularly in relation to process safety.

Last word

So to finish where we began, we agree with the commentators who claimed that the bonus structure at Texas City constituted a systematic disincentive to attend to process safety. But we would want to qualify the recommendations of the Baker panel reported in Chapter 1. We endorse the idea that refinery managers be held accountable for process safety indicators in their performance agreements. But we do not agree that the pay of lower level managers and non-managerial workers should be made to depend on these indicators. Nor should their pay depend on injury statistics. Lower level personnel should be provided with incentives to report bad news, rather than to suppress it.

Notes

1   A form of investment common to many UK derived legal systems. I am indebted to Kerry Jacobs, The Australian National University, for this idea.

2   Kerry Jacobs. Personal communication. The reference goes back to Beatrice Webb who talked of public servants as a Jesuitical corps of ascetic zealots (Hood, 1995, p. 94).

3 The unit trust proposal has been outlined here with minimal detail and will no doubt meet with scepticism from some readers. We invite the sceptics to think constructively about how such a scheme might be made to work.

4 Haigh, 2006, p. 22.

5 In many companies we visited the bonus and broader performance management systems were intertwined. As such, the performance evaluation not only affected the bonus, but the award of salary increments. Some informants raised the award of a salary increment as an important motivator for them.

6 Reason, 1997, Chapter 9.

Appendix 1
The Research Process

This book has been a long time in the making. Hopkins has been researching and consulting in the area of catastrophic risk for many years, and developed a particular interest in the impact of bonuses during his analysis of the BP Texas City refinery disaster of 2005. His accident analyses and consultancy observations incubated the general arguments about human motivation and catastrophic risk management, and ultimately, the development of a series of research questions for systematic inquiry. Maslen joined the research at this point and, together, we asked: What incentive arrangements are in use? To what extent are these targeting catastrophic risk management? And what effect do they have on employees anyway?

To shed some light on these questions, we took a case study approach. Eleven companies participated, whose operations included oil and gas, petrochemical, mining and pipelines. These companies were selected purposefully to provide samples from multiple hazardous industries, including companies who we identified as innovating in this area. We were after 'best practice'. The companies were mostly large multinational concerns, though a couple were smaller, operating in a single country. Most companies were publicly listed. There were also a few that were subsidiaries of larger, multinational operations.

Our study involved document analysis, as well as interviews and observation. Public and internal documents that detailed companies' performance agreements and incentive structures were one data source for this research. Annual, remuneration, and sustainability reports gave high-level information about remuneration structures, incentive metrics, metric weightings and rationale behind schemes for the CEO and most senior executives. We also collected internal documents on the performance management and incentive schemes. Not all companies gave access to this documentation, but where it was given, it gave a much more detailed look at the metrics and the mechanics of schemes.

Between us, we formally interviewed more than 40 people from these companies. To start with, we spoke with company representatives who explained the design, rationale and function of their incentive schemes. These discussions were had in the context of the documentation that we reviewed, and offered an opportunity for further detail and clarification on the components of the short- and long-term incentives, the weighting of different components, and the inclusion of safety metrics. Representatives included general managers, human resources managers and safety managers.

In the first stage of the research, we were unable to answer the vital question of the extent to which incentives impacted daily decisions. For this reason, we conducted a second stage of interviews in which we spoke with senior managers in a small number of the case study companies about what drives them at work, and the extent to which their bonus impacts their daily decisions. We asked about their role, the pressures they work under, their performance agreements, their performance evaluations, and the impact of the financial bonus in hypothetical and real terms. This stage included further document analysis, as we reviewed the performance agreements of participating senior managers. All interviews were recorded and later transcribed, with the consent of the participants.

In addition to these formal interviews, the research is informed by informal interactions and our observations. We sat in on meetings, were shown around facilities, and had countless conversations. This component of the research provided powerful insight into the work environment and the role of bonuses.

Appendix 2
A Sample Performance Agreement

Performance Agreement

Employee Information			
Employee Name:	**John Smith**	Employee ID:	**52406**
Job Title:	**General Manager**	Date:	**22/12/2012**
Department:	**Production Projects**		
Manager:	**Michael Anderson**		
Review Period:	**01/01/2013 to 31/12/2014**		

General

Set strategies to address long-term problems, identify future work needs, and negotiate resources. Takes into account economic, political, community, and business complexities. Deliver value for the Group.

Contribute to delivery of Business Objectives by carrying out the Scope of Work in line with the UNITY model.

Ensure compliance with standards and Group governance and assurance policies.

Represent the project to the external stakeholder community.

Promote and implement UNITY behaviours model across the business.

Demonstrate personal & visible leadership commitment to, and accountability for, safety.

Development and implementation of a consistent approach to project delivery across the business in line with the agreed strategy.

Responsible for delivery of Project A scope in line with schedule and budget as defined by the board.

Ensure project activities support long-term production and operability targets.

Ensure projects are within +/- 10% of cost and schedule, and liaise with Group projects team.

Top quartile project performance in line with company and industry benchmarks.

Lead discussions with the Project Board as required.

Effective risk management to mitigate delays and overruns.

Ensure roles, responsibilities, and accountabilities are understood and aligned with project objectives.

Provide leadership, mentoring, guidance, and support with leadership team.

Prepare company submissions to Board as required.

Ensure systems and processes are in place, consistent with Group systems.

Own the stakeholder plan for Project D, building effective relationships with multiple public and private bodies. Leverage and engage multiple internal players to ensure engagement and relationships are built at the most appropriate level.

Participate in sharing Best Practice in the business.

Ensure effective succession management across projects and support professional development.

Ensure Smature contractor management systems are in place.

Lead preparation of key approvals e.g. contract awards, project approvals

Key Objectives

Objective	Details	Weighting
Business KPIs	FINANCIAL KPIs - Total capital expenditure US$ xxx - Exploration expenditure US$ xxx - Cash flow US$ xxx - Deliver cost reductions US$ xxx - Opex US$ xxx - Capex US$ xxx NON-FINANCIAL KPIs - Net production xxxxx - Production efficiency %xx - Maintain project team focus for completion in 2014 - Support initiatives for operations readiness - Maintain relations with key community stakeholders	40
Project Milestones	Award Project A Stage 2 contract – Jan 14 Award Project A Stage 3 contract – Jan 14 Project A Stage 2 – Jun 14 Project Stage 3 – Aug 14 Project A complete and handover to operations – Oct 14 Project B complete and handover to operations – Nov 14 Commission Project C Phase 2 – Sep 14 Project D complete and handover to operations – Nov 14 Project E complete and handover to operations – Dec 14	35
Health and Safety	TRCF 0.0 Fatalities 0 Implement safety improvement plan Promote asset wide safety culture Operational safety cases in place Conduct 3 management asset safety visits	15
Staff management	Recruit and retain key positions Provide support to Project D team Conduct 2 general staff meetings Implement new UNITY model Maintain relationships with leadership team	10

Evaluation

Reviewer comments:

Reviewee comments:

Bibliography

Amabile, T., 1996. *Creativity in Context*. Boulder, CO: Westview Press.

Ariely, D., 2009. *Predictably Irrational: The Hidden Forces that Shape Our Decisions*. London: HarperCollins.

Bakan, T., 2004. *The Corporation: The Pathological Pursuit of Profit and Power*. London: Constable & Robinson.

Bebchuk, L. and Fried, J., 2010. *Paying for Long-Term Performance*, Discussion Paper 658 (2010) Olin Center for Law, Economics, and Business, http://www.law.harvard.edu/programs/olin_center/.

Bergin, T., 2011. *Spills and Spin: The Inside Story of BP*. London: Random House.

Deci, E., Koestner, R. and Ryan, R., 2001. "Extrinsic Rewards and Intrinsic Motivation in Education: Reconsidered Once Again", *Review of Educational Research*, 71(1): 1–27.

Diener, D., 1999. *Introduction to Well Control*. Austin TX: University of Texas.

Doran, G., 1981. "There's a SMART Way to Write Management's Goals and Objectives", *Management Review*, 70(11): 35–36.

Elton, L., 2000. "The UK Research Assessment Exercise: Unintended Consequences", *Higher Education Quarterly*, 54(3): 274–283.

Etzioni, A., 1988. *The Moral Dimension: Towards a New Economics*. New York: The Free Press.

Flemming, N., 2011. "The Bonus Myth", *New Scientist*, 210(2807): 40–43.

Goddard, M., Davies, H., Dawson, D., Mannion, R. and McInnes, F., 2002. "Clinical Performance Measurement Part 2: Avoiding Pitfalls", *Journal of Royal Society of Medicine*, 95: 549–551.

Haigh, G., 2006. *Asbestos House: The Secret History of James Hardie Industries*. Melbourne: Scribe.

Hayes, J., 2013. *Operational Decision-Making in High-hazard Organizations: Drawing a Line in the Sand*. Farnham: Ashgate.

Hood, C., 1995. "The New Public Management in the 1980s: Variations on a Theme", *Accounting, Organisations and Society*, 20(2/3): 94.

Hopkins, A., 2007. *Lessons from Gretley: Mindful Leadership and the Law.* Sydney: CCH Australia.

Hopkins, A., 2008. *Failure to Learn: The Texas City Refinery Disaster.* Sydney: CCH Australia.

Hopkins, A., 2009. *Learning from High Reliability Organisations.* Sydney: CCH Australia.

Hopkins A., 2009. "Thinking about Process Safety Indicators", *Safety Science*, 47(4): 460–465.

Hopkins, A., 2012. *Disastrous Decisions: The Human and Organisational Causes of the Gulf of Mexico Blowout.* Sydney: CCH Australia.

Mannion, R. and Braithwaite, J., 2012. "Unintended Consequences of Performance Measurement in Health Care: 20 Salutary Lessons from the English National Health Service", *Internal Medicine Journal*, 42(5): 569–574.

Pink, D.H., 2009. *Drive: The Surprising Truth About What Motivates Us.* New York: Riverhead Books.

Porter, T., 2009. "The Division of Moral Labour and the Basic Structure Restriction", *Politics Philosophy Economics*, 8(2): 173–199.

Putnam, R., 1993. *Making Democracy Work: Civic Traditions in Modern Italy.* Princeton NJ: Princeton University Press.

Reason, J., 1997. *Managing the Risks of Organizational Accidents* Aldershot: Ashgate.

Schein, E.H., 1997. *Organizational Culture and Leadership.* San Francisco: Jossey-Bass.

Weick, K.E. and Roberts, K.H., 1993. "Collective Mind in Organizations: Heedful Interrelating on Flight Decks," *Administrative Science Quarterly*, 38(3): 357–381.

Werner, R. and Asch, D., 2005. "The Unintended Consequences of Publicly Reporting Quality Information", *Journal of the American Medical Association*, 293(10): 1239–1244.

Yeh, S., 2010. "Financial Sector Incentives, Bailouts, Moral Hazard, Systemic Risk, and Reforms", *Risk, Hazards & Crisis in Public Policy*, 1(2): 97–130.

Index

References to illustrations and diagrams are in **bold**.